David Fernandez Rivas
Empathic Entrepreneurial Engineering

Also of Interest

Technology Development
Lessons from Industrial Chemistry and Process Science
Ron Stites, 2022
ISBN 978-3-11-045171-9, e-ISBN (PDF) 978-3-11-045163-4

Engineering Innovation
From idea to market through concepts and case studies
Benjamin M. Legum, Amber R. Stiles, Jennifer L. Vondran, 2019
ISBN 978-3-11-052101-6, e-ISBN (PDF) 978-3-11-052190-0

Product-Driven Process Design
From Molecule to Enterprise
Edwin Zondervan, Cristhian Almeida-Rivera, Kyle Vincent Camarda, 2020
ISBN 978-3-11-057011-3, e-ISBN (PDF) 978-3-11-057013-7

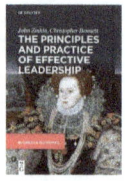
The Principles and Practice of Effective Leadership
John Zinkin, Christopher Bennett, 2021
ISBN 978-3-11-070780-9, e-ISBN (PDF) 978-3-11-070787-8

Scientific Leadership
J. W. (Hans) Niemantsverdriet, Jan-Karel Felderhof, 2018
ISBN 978-3-11-046888-5, e-ISBN (PDF) 978-3-11-046889-2

David Fernandez Rivas

Empathic Entrepreneurial Engineering

—

The Missing Ingredient

DE GRUYTER

Author
Prof. David Fernandez Rivas
Faculty of Science and Technology,
Mesoscale Chemical Systems
University of Twente
Drienerlolaan 5
7500 AE Enschede
Netherlands
d.fernandezrivas@utwente.nl;
www.david-fernandez-rivas.com

ISBN 978-3-11-074662-4
e-ISBN (PDF) 978-3-11-074682-2
e-ISBN (EPUB) 978-3-11-074690-7

Library of Congress Control Number: 2022932966

Bibliographic information published by the Deutsche Nationalbibliothek
The Deutsche Nationalbibliothek lists this publication in the Deutsche Nationalbibliografie; detailed bibliographic data are available on the Internet at http://dnb.dnb.de.

© 2022 Walter de Gruyter GmbH, Berlin/Boston
Cover image: RichVintage / E+ / Gettyimages
Typesetting: VTeX UAB, Lithuania
Printing and binding: CPI books GmbH, Leck

www.degruyter.com

Contents

Foreword by Robert S. Langer —— XIII

Foreword by Jarka Glassey —— XV

Preface —— XVII

Acknowledgments —— XXIII

1   I am short of time! —— 1

2   Where to start engineering? —— 3
2.1   Overall aim of this book —— 3
2.2   Boundary conditions or book's aim —— 4
2.3   Conceptualizing terms —— 6
2.3.1   Engineer, E_3 —— 6
2.3.2   Entrepreneur, E_2 —— 8
2.3.3   Knowledge, K —— 9
2.3.4   Persuasiveness, P —— 10
2.3.5   Empathy, E —— 13
2.4   Mixing the ingredients: K+P+E —— 16
2.5   KPE against epidemics —— 20
2.6   Applicability of ideas —— 24
2.7   A decision making tool —— 25
2.7.1   Origins of IF —— 31
2.8   Relation between A and *IF* —— 32
2.9   Changing K:P:E importance —— 33
2.10   KPE's model case —— 35

3   Give me facts ... —— 37
3.1   Innovation guide —— 37
3.2   Inspiration template to innovate —— 42
3.3   Size matters —— 45
3.4   First-hand case analyses —— 47
3.4.1   From PhD student to entrepreneur —— 47
3.4.2   Starting another company —— 57

4   Where did I read this? —— 62
4.1   Overall learning objectives of this book —— 62
4.2   A new engineer —— 64
4.3   Drivers for paradigm change —— 67

4.4	Method to teach KPE —— 69	
4.4.1	Learning from challenges —— 69	
4.4.2	Teaching knowledge —— 71	
4.4.3	Teaching persuasiveness —— 73	
4.4.4	Teaching empathy —— 76	
4.5	The need for teaching empathy and persuasiveness —— 77	
4.6	How others teach entrepreneurship, E_2 —— 79	
4.7	Four recent books —— 80	
4.7.1	Book 1. Tropical Empathy —— 80	
4.7.2	Book 2. Entrepreneurial physicists —— 81	
4.7.3	Book 3. Hunch Engineers —— 82	
4.7.4	Book 4. Chemistry Entrepreneurs —— 83	
4.8	Solving changing problems —— 83	
5	Got it! Now what? —— 87	
5.1	Take-home message —— 88	
5.2	KPE^3 tips —— 88	
5.3	Less academic take on persuasion and empathy —— 90	
5.3.1	Persuasiveness vs propaganda —— 90	
5.3.2	Empathy vs apathy —— 92	
5.4	Some advice and biases —— 94	
5.4.1	Off- and online pointers —— 96	
5.5	Communication etiquette —— 97	
5.5.1	Writing tips —— 98	
5.6	Models and dialogues with yourself —— 99	
5.7	To lead or to follow? —— 104	
5.7.1	About leaders and followers —— 104	
5.8	Watch out! —— 106	
5.9	About dog-eat-dog —— 110	
5.10	About teams —— 111	
5.11	Success or failure, what is the question? —— 113	
5.11.1	The value of time —— 114	
6	IEEE: Interviewing empathic entrepreneurial engineers —— 118	
6.1	Empathy-driven need to write —— 119	
6.1.1	Distilling ideas —— 119	
6.2	KPE seen by other innovators —— 121	
6.2.1	Artificial intelligence and chemistry —— 121	
6.2.2	A is not always for Apple —— 125	
6.2.3	Swab sensor —— 127	
6.2.4	Chemistry is in the air —— 129	
6.2.5	The challenges are there, but can we see them? —— 132	

6.2.6	Flying droplets —— 133	
7	**Epilogue: Last considerations —— 136**	
A	**Answer to case study questions —— 139**	
A.1	Knowledge answer —— 139	
A.2	Persuasiveness answer —— 141	
A.3	Empathy answer —— 143	
A.3.1	Answer 1 —— 144	
A.3.2	Answer 2 —— 145	
A.4	*IF* cleaning answer —— 146	
A.5	Examples used in lectures to teach *IF* —— 146	
A.6	Oscillatory baffle reactor —— 147	
A.7	Organometallic reaction in fine chemical and pharmaceutical industry —— 148	
A.8	Answers to the rhetoric question from Section 2.3.2 —— 150	

Bibliography —— 151

Index —— 159

To my children because they will see the future
and to my family who helped me get where I am

"Ciencia y libertad son llaves maestras que han abierto las puertas por donde entran los hombres a torrentes, enamorados del mundo venidero."

Science and freedom are the master keys that have opened the doors through which men pour in, falling in love with the upcoming world.

José Martí, "Respeto a Nuestra América," *La América*, New York, 1883

Foreword by Robert S. Langer

It is an honor and a pleasure to write a foreword to *Empathic Entrepreneurial Engineers* by David Fernandez Rivas. This is a terrific book that combines case studies, interviews with entrepreneurs, and other relevant pedagogical tools. It is a book that should be very useful for teachers who will use it in both undergraduate and graduate programs. This book also makes a major effort to nurture empathy to everyone who reads it, though it is largely aimed at STEM students and professionals.

Empathy is considered as one of the durable skills that are needed to prepare the professionals who will face the challenges of the 21st century. Durable skills – so-called "soft" or "people-related" skills – are meant to balance the toolbox engineers and other professionals should have. These are particularly handy for interdisciplinary teamwork. Teaching durable skills and nurturing entrepreneurial character in students and professionals is a laudable challenge that not many educators or leaders tackle.

This book should serve as an innovation guide for both people new to innovation and experienced professionals. David gives the reader, through conversations and short interviews, a feel for what different engineers do and think and how their companies develop. The journey these innovators take should provide inspiration for your own journey.

David brings an interesting combination of elements that respond to his professional trajectory, which he shares in his book. He has shown an outstanding commitment to research that advances science and technology targeted at sustainable solutions to global societal challenges. His talent is in connecting fields and people as shown, for example, by the many co-authors of his published work in very broad subject areas ranging from physics, chemistry, chemical engineering, and materials, energy, and environmental sciences to bioengineering, education, and social sciences.

With his home base now at the University of Twente, The Netherlands, he has an excellent publication record, including research done in Cuba under difficult circumstances. He has had research visits at numerous top institutions, including the Massachusetts Institute of Technology (MIT), where he is Research Affiliate with the Mechanical Engineering Department, École Polytechnique Fédérale de Lausanne (EPFL), the University of Melbourne, Dortmund University, and the International Centre for Theoretical Physics in Trieste.

David was chosen as the Dutch Engineer of 2021 by the Royal Dutch Institute of Engineers, showcasing his devotion as scientist with an open eye for real-world applications. I find it remarkable to see how he balances excellent scientific research with well-planned and visible efforts to valorize it. He communicates enthusiastically with the public media and creates companies and products used in society.

After receiving his degree as a nuclear engineer, David has grown out to the larger length-scales of our physical world, adapting to new areas. His efforts align well with sustainable development goals, such as Good Health and Well Being, Quality Education, and Clean Water and Sanitation. One of his ambitions that stands out due to the relevance and urgency it has today is the development of needle-free injections for biomedical applications such as vaccination or insulin treatment.

Finally, David has a quality that is especially vital these days: He is a good communicator. In this book, as in his regular work, he often uses art and humor to get his message across. He has innovated low-cost technologies that will help improve the quality of life in both developed and developing countries. Doing all this matches the title of his book *Empathic Entrepreneurial Engineering*. I think you will very much enjoy reading it.

Robert S. Langer, Sc. D.
Institute Professor
Massachusetts Institute of Technology

Foreword by Jarka Glassey

Books like *Empathic Entrepreneurial Engineering* by David Fernandez Rivas do not come around often, but this combination of professional skills is particularly timely as we face some of the greatest societal challenges in human history. Whilst the technical competence of engineers of the future cannot be underestimated, and their understanding of fundamental principles of engineering will be essential in developing safe solutions, they will surely help us ensure healthy and fulfilling lives for people without killing our planet in the process. However, they are by no means going to be sufficient.

Innovation and entrepreneurship as well as empathy will be needed to make sure the proposed solutions make the necessary radical steps in the way we manufacture products and provide services for those who need them. Ethical and responsible engineering is rapidly gaining prominence worldwide as we recognize that digitalization of industries and the rapid rate of scientific discoveries potentially raising ethical questions about their use as well as the much-needed transitioning to a low-carbon circular economy will demand much more from our engineers than just technical excellence and competence.

However, engineering educators would often tell you that teaching these "durable" skills, as David refers to them, in the engineering context is very difficult. David shows in his book you are about to read that it is not only critical, but it can also be fun and exciting, and it can bring many more, diverse future engineers to the profession. This book should prove an invaluable resource for engineering educators, students, and engineering practitioners alike. But his use of case studies, interviews with entrepreneurs, and examples of exciting developments in engineering should be very approachable for future budding engineers that never saw engineering as a career option.

The clear explanations of new pedagogical concepts and paradigms David proposes are also a valuable underpinning to new engineering pedagogy that will help us improve the delivery and the assessment of engineering education in the future.

I have had the pleasure of working with David on engineering education for years now and his passion and enthusiasm for doing the right thing and doing it well not just in his research, but also in educating future generations of engineers always impressed me. I am sure that once you read this book, you will agree with me that David's passion and clarity of explanation make this new paradigm of Knowledge, Persuasiveness, and Empathy (KPE) such an obvious approach that you will wonder why it has not been introduced before. Are you wondering what KPE paradigm I am referring to? Well, read on and enjoy!

Jarka Glassey
Professor of Chemical Engineering Education
Newcastle University, UK

Preface

Thank you, reader, for giving me the chance to present to you this book.

The urgent message this book carries is a request to nurture *empathy* in all the people we can reach.

Why empathy?

The most moving interaction I have had in the last 10 years was with parents of small children fighting Type I diabetes, and the young adults who show so much bravery in their daily activities, constantly pricking their skin and injecting with needles the insulin and glucagon they need to stay alive.

Understanding the actual limitations of state-of-the-art technology was only possible after I talked to these people who really need a better solution. The information I gained with those conversations is not plainly written in the scientific journals or technological reports. You will not easily get it from the doctors or medical companies that have developed the best technologies.

Empathy enabled me to align very fast all my efforts towards the needs of the people who will benefit from the technology my team and I are developing, particularly to inject without needles.

Our world is in desperate need for other solutions to the many problems we face as humans, and I strongly believe you have chances of contributing positively to the whole challenge.

Why now?

Like my best teachers, I want to start with some questions, but please try to think about your answers before reading mine.

---

1. What if we could have a framework, independent of which historical moments we are living, that helps guiding in any direction: technical, economical, commercial, ecological, legal, etc.?
2. Could we strip down to the bare minimum all those concepts and great books that I have seen to build such timeless framework?
3. Would such a minimalistic approach be useful for someone out there?

Answers:
1. I am presenting you with an updated definition of engineering, knowledge, persuasiveness and empathy.
2. I compiled a nonexhaustive list of advice or tips for the future you may be facing anytime soon. It is this strong feeling that such guide is missing and necessary that motivated me to prioritize my time to work on this book.

3. I believe the answer is yes, based on the feedback given by my collaborators and former and current students. However, the ultimate answer will be if students and professionals can apply in their lives the things they learn from this book.

Who am I writing for?

Have you ever felt that something wrong around you could or should be fixed? For example, reducing the gas emissions of public transportation means, say a passenger bus. Do you feel that you are unprepared, or do you get frustrated that there is no clear path or point from where to start, such as how to get in contact with the bus manufacturer?

Even if you have spent years studying or working – at home or employed – I think that there is nobody who could resolutely say there is a magic formula or a step-by-step guide to solve a given problem. When I say "solve a problem" it may be something small, like solving a simple math problem, but it can also apply to something (really) big, such as trying to solve even a small aspect of the climate crisis.

I see this book as a sort of innovation guide for newbies and experienced professionals. If you are reading this as part of a course it will help those who feel the need to do more than just get another academic grade or degree and want to contribute to solving real-life problems.

This book will ...
- provide you with my humble "cocktail" of basic ingredients that I think are needed for you to get started solving problems: knowledge, persuasiveness and empathy (KPE);
- clarify the words innovation and entrepreneurship, which you might have heard in different contexts;
- trigger very different ideas or feelings when discussing about soft skills, depending on your personal and professional journey;[1]
- fill a knowledge gap: a "how-to-start-innovating guide for newbies" who are facing the twenty-first century;
- teach you new perspectives that expand beyond stories of people who already succeeded many years ago, where I refer to success in the context of achieving solutions and personal growth;
- give you a temporal proximity to "solutions in the making" explained by several innovating engineers and their companies. Their journeys may give some inspiration for your own journey.

---

[1] The challenge to teach students "how to innovate" for several years gave me the springboard to start writing about this topic.

This book will not ...
- necessarily be the best starting point if you already have a clear solution and aim to start a company or entrepreneurial activities; however, you can check Section 4.6 and Chapters 5 and 6 for useful tips;
- be one of the following two types of books or articles typically found:
   (a) containing only anecdotal stories of famous inventors, leaders of companies and politicians, labeled "a success" a long time ago, and
   (b) very technical, academic or philosophical, which are mostly helpful for specialists in specific activities, such as product development, humanistic studies, business management, etc.;
- replace any of the available literature. It will be complementary to the types of literature listed above, a guide to action to get you started solving problems or at least trying to do so.

Thus, this book is meant for anyone who:
- wants to *know more or teach* about solving problems regardless of background knowledge;
- wants to undertake an innovation effort to improve or create a product or a service, take a scientific challenge, etc.;
- is considering creating or joining a company to become an active agent of change,
   - ... or is just undecided; that's also fine;
- learn about other people's experience in doing all of the above.

Whatever your reason might be, I am very interested in establishing a dialogue with you, the reader. My aim is to improve this first attempt I have made to teach my tricks to new generations of problem solvers.
   For this, please join the discussion at www.empathic-engineering.com.

In practice, this book is in the first place meant to be used in or along undergrad science, technology, engineering and mathematics (STEM) programs, i.e., BSc and MSc, but written in a way that can be understood by other people.
   The majority of the examples are related to chemical engineering, with a focus on microfluidics and process intensification, because these are my most active areas of research. However, they are discussed at a basic level so that it can be understood by students without any prior chemistry-related knowledge.

Why am I the right person to tell you about this?

My mission with this book is to help STEM students primarily to be entrepreneurial and empathic. If you, as reader, consider to already have some of these qualities, bear with me because I believe we can always have a bit more of both qualities.
   I was implicitly trained to innovate as a primary task (engineering, some would say). Yet, even when many study programs include innovation in one way or another,

students focus more on specific (technical) subjects based on knowledge you can find in classical textbooks or biographies – the trend changes slightly after graduation in real life.[2] Yet, I felt that the newer generations might need something extra that should not overload the already cramped study plans.

To avoid the trap that this book becomes another innovation or management guide,[3] I tried to isolate myself as well as I could from what was already written, at least at the beginning. This allowed me to think how my past experiences could help those who, like me, would like to change the world for the better.

Reading any of the books and internet sources listing anecdotes or examples from innovators may not necessarily show you how to innovate, because it is hard to reproduce situations from other places and contexts – particularly if they happened long ago. However, I found analogies and convergence points between my ideas and the literature, which is a good sign, but I could not find a reasonably compact guide to innovate "right away."

I want this book to help you while you are actually trying to solve the problem of your choice, now or in the future. To satisfy this desire, I spent some time packing different ideas, stripping them to the bare minimum, so that you can easily remember the basics.

My interest in getting all this out as soon as possible was fueled by the unique circumstances our world has endured since 2020. With the lockdowns imposed around the globe to reduce the impact of COVID-19, I could only imagine how people felt restricted to their rooms or houses, unable to study or feel useful. I tell more about this in Section 6.1.

In particular, my daughter was supposed to experience her first year at university. I started thinking, if I were in her situation, or at least her age, "what would I have done?" Empathizing with her, my students and other people limited by the pandemic, I searched my memory and published work to distil this book for you.

How to read it?

This book is not meant for you to read it in one go. If it is not used in a course, where your teacher will set the order and pace, I envisaged you could use it more like a consulting text or an entertainment non-fiction book. By that, I mean that you may benefit from reading chapters and sections in the proposed order, but you may also start where

---

[2] One of my students said to me: *"you have the unique advantage of not only having tried to innovate, but also having grappled with the challenge of trying to teach innovation. This sets your book apart from the in vogue entrepreneurship books."*

[3] There are many great books that serve these purposes! Some of them I duly cite.

you have the most urging question. There is sufficient cross-referencing to bring you from one side to the other.

I am naturally inclined to suggest you to try and start, er ..., from the beginning! The main reason is that it starts "light" in content, and when it gets more dense, you will get warnings of a "rabbit hole" ahead – thus, enter if you have time, who knows what you might learn inside. The benefits of entering such holes tend to be evident in the long term.

Anyway, the chapters are:
1. Chapter 1, I am short of time! *Everyone* — page 1
2. Chapter 2, Where to start engineering? *Students/teachers* — page 3
3. Chapter 3, Give me facts.... *Everyone* — page 37
4. Chapter 4, Where did I read this? *Teachers and professionals* — page 62
5. Chapter 5, Got it! Now what? *Everyone* — page 87
6. Chapter 6, IEEE: interviewing empathic entrepreneurial engineers. *Everyone* — page 118

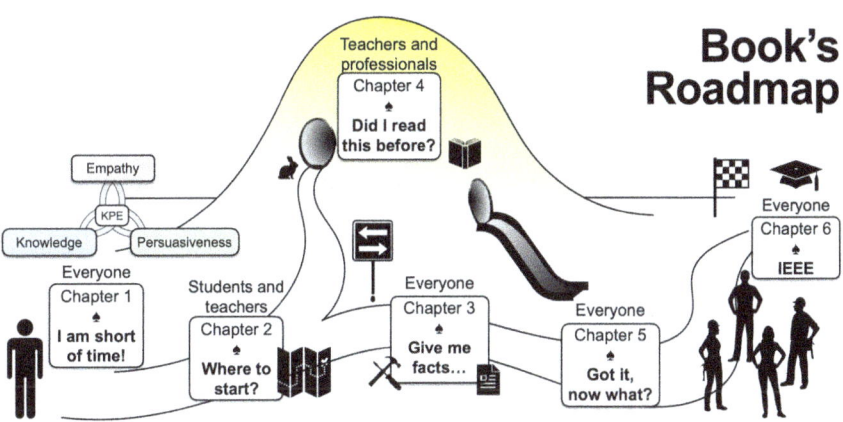

Figure 1: Suggested guide to read this book. Chapter 1 is so short that it would be a pity not to read it. Chapter 2 is still easy to grasp, while Chapter 3 gives more details and useful tools. Chapter 4 is a bit tougher to digest, and I put a rabbit hole to represent that it can get deep into some aspects. However, if you are in a rush, skip it for now, and go on to Chapter 5. Lastly, Chapter 6 gives snapshots of personal journeys through interviewing empathic entrepreneurial engineers.

I trust the reader will identify a logic structure in the choice of chapter titles, which I briefly explain now.

I know how difficult it is to keep our attention or focus on a single task or objective these days: everyone seems to have very limited time for anything. I condensed the most important elements of this book to fit in Chapter 1, which is only one page, not counting its Figure. I give here what you really must know to start solving problems with "my ingredients" composing the KPE framework. I hope just glancing at it will trigger your curiosity and you will go further.

Chapter 2 gives a bit of guidance and expands on the basic elements given in Chapter 1, with basic tools that can help in the journey. Of great relevance, I provide my own definitions of what engineering should be today, or at least in the near future. In Chapter 3, my ambition is to give you a few examples, mainly from my own experience. This chapter may be updated or grow with the passage of time, e. g. newer editions.

Chapter 4 is for those who want to have more references to dive into or check what others have done in the past. Here I want to clarify the question *"Where did I read this before?"* citing most of the relevant books or articles I used, and that can help the interested reader go on further. I may have missed relevant publications. Therefore, send me a message if you can, suggesting any you believe should be included.

Chapter 5 will have a coaching character by providing tips aligned with the KPE framework that help in making steps to connect effectively to other people and grow professionally on your own. In Chapter 6, my views are analyzed through the lens of other innovators I know.

Along the way in the book, you will also find notes, assignments and cases with a strong link with real-life scenarios, besides personal experiences of colleagues and myself. The answers or extra evidence backing up the statements in the main text can be found in the appendices at the end of the book. I did this with the intention of clarifying the more abstract ideas discussed along the book's narrative line.

These side steps in the story will help you apply knowledge for solutions, identify points of attention for your own self-assessment and find sources of inspiration.

Acknowledgments

I am very grateful to several people and institutions that have allowed me to gain access to material I used for this book. I want to begin thanking the University of Twente, which has been my home-base, and all the support staff and colleagues since I joined in 2007. I want to thank the MESA+ Institute, NovelT, and the DesignLab for facilitating new initiatives and innovations such as The Future Under Our Skin (FUOS) [2]. I am privileged to have been supported also by the Mechanical Engineering Department and the Edgerton Center at the Massachusetts Institute of Technology (MIT), USA. The corrections for this book were done during my sabbatical at the École polytechnique fédérale de Lausanne (EPFL), Switzerland, in the spring of 2022.

The team at De Gruyter has been a unique source of support, particularly Karin Sora who believed in my dream of writing this book and made it happen with her team: Ute Skambraks, Ina Talandienė, Vilma Vaičeliūnienė. Volker Hessel was the one who put me in contact with Karin, and that deserves a big 'Thank you'!

A special 'thank you' goes to the people who made possible to have a unique prologue for this book.
- Robert Langer kindly agreed to write about the practical aspects of this book. I also thank Samir Mitragotri who put us in contact.
- Jarka Glassey has given a complimentary analysis of what this book can provide to the future engineers and their teachers. I met Jarka thanks to my long-time friend and collaborator, Daria Camilla Boffito.

In several stages, I counted with special support from Gareth McKinley, Joe Niemela, Ian Hunter, and Lynette Jones, who always gave extremely good ideas that helped me improve the quality of this book.

My gratitude goes also to several funding agencies and organisations that have provided support and a platform to amplify my message over the years: the Dutch Research Council (Nederlandse Organisatie voor Wetenschappelijk Onderzoek NWO), the Dutch Research Agenda (NWA), The Royal Holland Society of Sciences and Humanities (Koninklijke Hollandsche Maatschappij der Wetenschappen KHMW), The Royal Netherlands Society of Engineers (Koninklijk Instituut Van Ingenieurs KIVI), the European Research Council (ERC), The Abdus Salam International Centre for Theoretical Physics (ICTP), the Young Academy Europe (YAE), the Global Young Academy (GYA), and the magazines *U-Today*, *De Ingenieur* and *The Chemical Engineer*.

Those who read this book in the early stages had to digest a half-baked idea, but the book was shorter. The ones reading the book towards the end of the project had more to read, but benefited from the first waves of support. In sum, writing this book would have been impossible without the support from the following people which I provide in a loosely established order:
- Sebastian Husein helped me with being critical about concepts I really needed a fresh angle. The "interesting entrepreneurial engineers" I interviewed in Chap-

ter 6: Miguel Modestino, Daniela Blanco, Connie Nshemereirwe, Richard Novak, Stafford W. Sheehan, Akash Raman, and Tom Kamperman. Surya Raghu, Dave Blivin, Bob Lansdorp and Gerard Cadafalch really helped me during conversations about entrepreneurship and their own company's journeys. Daniel Jitbahadoer and Arian Hohmann helped me a lot with design ideas and unconventional support!

– Han Gardeniers, Detlef Lohse, and Pedro Cintas Moreno have been like fathers since my PhD years, and have always been a sounding board;
– colleagues at the University of Twente, from different departments: Devaraj van der Meer, Albert van den Berg, Marike ter Maat, Kostas Nizamis, Frank van den Berg, Daniel Chehata, Chantal Scholten, Alvaro Marin, Mathieu Odijk and Tanya Bondarouk, provided very valuable insight at different stages, Nelly Litvak inspired me to focus on our STEM students;
– from other institutions, Davide Iannuzzi, Andrzej Stankiewicz, Andrew Dickerson, Javier García-Martínez, Robert Lepenies, Aldert Kamp, Alfo Batista, Javier Rodríguez, Alberto Levy, Luis Perez-Breva, Jealemy Galindo, David Tarbay, Ernesto Altshuller, Miguel Delcour, Eveline van Zeeland, Chris Moser and Jorge Parada;
– the work of science communicators and journalists was an invaluable source of tips: Laurens van der Velde, Michaela Nesvarova, Jim Heirbaut, Adam Duckett, and Jennifer Chu;
– from my team of grad and postdoctoral students: Diana van der Ven, Miguel Quetzeri, Keerthana Mohan, Dawid Surdeko, Jelle Schoppink, Pep Canyelles, Nicolás Rivera Bueno, and a very big thanks to 'the Dude' Stefan Schlautmann;
– other former students that were instrumental: Kevin Rouwenhorst, Maarten Ladrak, Bas Koelewijn, and Jochem Tijburg.

A very special 'thank you!' goes to my family and the support I have received throughout my life. Beginning with my mother, Pura María, who has inspired me since I was little. I still remember seeing her working at the Chemistry lab in one of the oldest breweries in Cuba. My fascination only grew when I found she wrote a book about her father, i. e., my grandfather, Virgilio Rivas,[4] a man who shaped my personality and still today I look up as example. My cousin Wendy Barnet, who has shared with me many moments and ambitions, namely the dream of writing books. She has published her first book for children and I look up to collaborate very soon in this endeavour.

Among my family members who could not enjoy with me the culmination of this book are two very special loved ones that require a separate mention. My father, Israel Evaristo, took me when I was a small boy to his work at the Editorial Científico-Técnica, in Havana. I remember with a bit of nostalgy the smell of fresh ink at the press while printing stories that I could not have imagined have defined what I do now: Science

---

4 Pura M. Rivas González. *Nosotros, el pueblo*. Editorial Orbe, Ciudad de la Habana, 1977.

and Technology. He also wrote a book, about mathematical puzzles,[5] and I am sure he would have been proud to see I kept the family tradition of writing books. Then comes my aunt, Mabel, who inspired me with her work attitude and unconditional support at every step I took. I will never forget the sound of the typing machine when she worked translating documents, and later when computers came into the picture, the careful way of reading and making sure the message was clear and accurate. She also wrote a book, which Wendy, her daughter is editing now.

Lastly, the time during which I wrote this book was a contrasting period in my life. It was full with very good and sad moments for the extended family, but more so for my closest or nuclear family, in particular Lea Milovich, my wife. She, her brother Lazar and her father Marko suffered the loss of my mother-in-law, Miriam to a disease that impeded her to meet her youngest grandson. Lea and our two sons, Leo and Daniel, have been with me 24/7 in this 'author-journey' and I could not imagine better companion. I experienced inspiring moments changing diapers and walking outside for fresh air with them. The geographical distance from my daughter Gabriela did not stop me from thinking how to make something useful for her, too. She was one of the first I pitched the book idea and I had her in my thoughts as a 'potential' reader for this book.

To all these special people, including you, the reader of this book: thank you for picking it up! I can only hope its content will keep inspiring you and future generations.

---

[5] Israel Fernandez Pujol. *Pasatiempos matemáticos*. Editorial Científico-Técnica, 1983.

1 I am short of time!

Reading time ~ 2 min

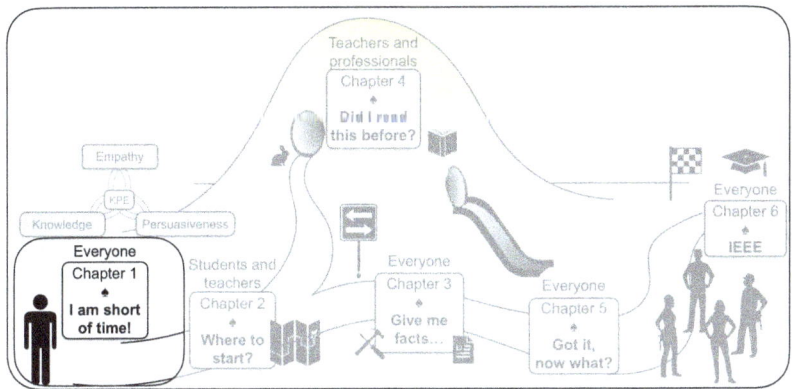

Bit: The overall idea of this book in one figure

"What is time? It would be nice if we could find a good definition of time. (...)
Perhaps we should say: 'Time is what happens when nothing else happens.' (...)
Maybe it is just as well if we face the fact that time is one of the things we probably cannot define (in the dictionary sense), and just say that it is what we already know it to be:
it is how long we wait!
What really matters anyway is not how we define time, but how we measure it."

Richard Feynman in The Feynman Lectures on Physics. Volume 1, 5–1 [99], and also here (https://www.feynmanlectures.caltech.edu/I_05.html).

Knowledge, persuasiveness and empathy (KPE) are the basic ingredients of the framework I propose as *a method to clarify problems and solve them*. This method is simpler than other alternatives, some of which will be discussed and compared in this book.

The main message of this book is to share the KPE framework and equip the engineers of the future with a multipurpose innovation tool.

KPE is a simplified conceptual framework based on three ingredients that are key for any engineer, no matter whether you want to design solutions for a problem, initiate a start-up or improve an existing organization as an employee.

KPE is universal because it is based on human qualities that barely changed in the last couple of centuries. I demonstrate it by analyzing several real-life cases and interviews given by empathic entrepreneurial engineers. A graphical representation of KPE and the six steps of the Innovation Guide are shown in Figure 1.1.

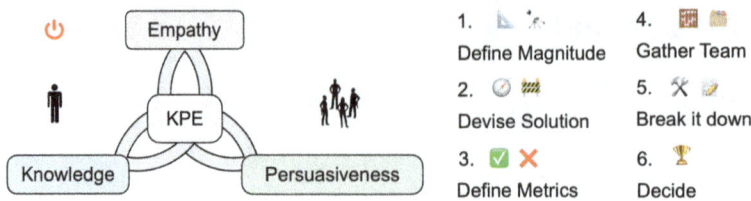

Figure 1.1: Summary of the most relevant ideas of the book. The interconnection between the KPE ingredients and their relation to the Innovation Guide steps will be unveiled in subsequent chapters, together with other engineering tools and tips.

Do not worry if you think you lack some KPE ingredients: you can acquire them while solving problems. Clarity about a problem is gained while exploring potential solutions that might never become a reality. There are several examples and tips in this book to help you understand this process.

When solving problems, you need to validate each assumption as fast as possible, for example by getting out of your usual workplace and asking any person you can reach out to: learn fast while failing.[1] Ask for help and get a team to avoid wasting time and resources. Avoid toxic environments or professional behaviors that may lead you off-track.

In sum, this chapter may help you decide to continue reading the book. At least you now have the backbone of my proposed way to solve problems of a universal character.

---

[1] See page 35.

2 Where to start engineering?

Reading time ~ 60 min

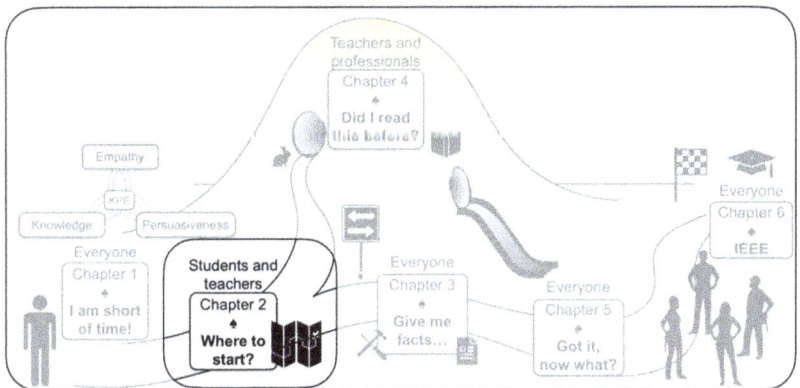

Bit: Important concepts and tools to help solving problems

> "If I have seen further it is by standing on the shoulders [sic] of Giants" –
> Used in different contexts, most attributed to Isaac Newton.
>
> (https://en.wikipedia.org/wiki/Standing_on_the_shoulders_of_giants)

I must warn you that some parts in this chapter might be philosophical and dry, but they are necessary for you to get the definitions and concepts used throughout the book. This chapter contains elements of my own work that are published in different peer-reviewed articles, and accordingly cited. In fact, the main idea behind this book is underscoring three core components or ingredients: knowledge, persuasiveness and empathy (*KPE*). KPE is a simplified conceptual framework for the training of innovators I started building in early 2020 – see more details in Sections 6.1 and 6.1.1 [96]. But before we dive into the actual way to solve problems, it makes sense to provide some boundary conditions and basic definitions.

2.1 Overall aim of this book

I want to help you to incorporate more empathy and entrepreneurship in your professional practice. I strongly believe it can increase the probabilities of you contributing positively to a sustainable and more humane world.

After reading this book, you will be able to be critical about yourself in your role as an engineer, teacher, researcher or any other professional avenue.

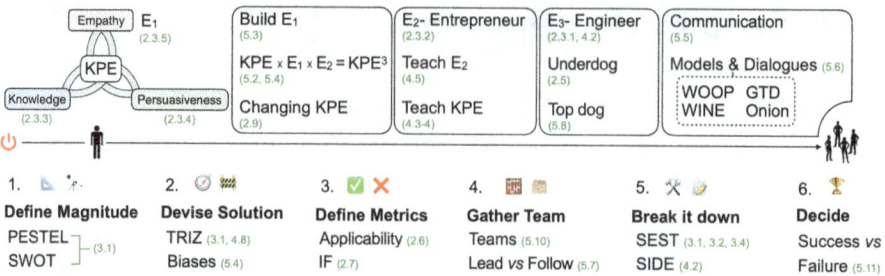

Figure 2.1: A graphical representation of the book, with the most relevant concepts, tips and tools and their corresponding sections (in green). The index "link to page" and the different sections will provide you with the explanation and meaning of all the acronyms. You can read this graphic from left to right, top-down, as a first approximation, but it does not follow exactly the order of chapters in the book.

To make this goal feasible and tangible for you, the start of each chapter lists the expected knowledge gain and the activities you should be able to execute after reading it. I give a more pedagogically oriented list of learning objectives aimed at teachers or people who are studying on their own in Section 4.1. To help you navigate this book, besides the classical index at the end, an overview is visualized in Figure 2.1.

2.2 Boundary conditions or book's aim

I have created some rules for myself while writing this book to achieve my goal of getting my message across. These rules will increase the chance that you apply what you learn in real-life scenarios, even when a few rules may challenge the expectations of some readers:

1. *I aimed to make the book just long enough.* I strongly believe that one can always stretch an idea, yet, very good ideas are best when short and simple.[1] I like using graphics or schematics to make my point in lectures or presentations. It saves a lot of time writing and we can grasp many concepts faster – assuming the schematics are clear enough. The downside is that they take some space on a page, thus I tried to keep the number of images at a minimum but without sacrificing the clarity and visual advantages.[2]
2. *It has to allow self-paced study.* Most of the elements I am providing have an origin in the work I have carried out in the last decade, peppered by short side steps based on other nonscientific readings I regularly do. Thus, this book is sometimes

---

[1] Think of the "elevator pitch," or when you have to explain in a couple of minutes what you are doing.
[2] The KISS principle and its many interpretations, e. g., Keep It Short and Simple [69].

nontechnical and pedagogical on purpose, so that it can be understood by any "layperson."[3]

3. *I will loosely follow formatting guidelines* regarding the number of words and chapters. This is necessary since I am trying to imagine all the different settings you may be in while reading any of the sections in the book. Human activity nowadays has a rhythm hard to follow for many. Not too many people have "time" to read in ideal circumstances (comfortable chair, sipping some beverage and meditating about each relevant passage).

4. *For younger generations* of students and professionals, the pressure to learn ever growing facts or assimilate information is much higher than for any previous generation. Therefore, I have made the effort to mimic the way information is consumed as I write these pages. At the beginning of each chapter, you will have a short take-home message, or *bit*, not exceeding 280 characters.
*Each **chapter** will show the approximate time it takes to read.*

5. *I give credit* by providing references to where I have taken ideas that are not originally mine, mostly in Chapter 4. But to make it comply with Rules 1 and 2, and not make it "yet another long textbook," I had to leave out some. As a disclaimer, there might also be cases where I do not recall anymore which source put me on a path, or perhaps something very similar has been published and I was not aware of it. In my defense, as Gribbin has pointed out, the history of science and technology is full of examples where an idea emerged in different places without direct influence [104].

6. The last rule is to *tell you all I can teach you in a more personal approach*. That is not common for textbooks, I know, but most of what I am sharing is based on properly cited objective tools and facts. Moreover, I clearly state my impressions or explain my subjective analysis. This will serve the following main purposes:
    (a) You will not feel too "distant" from your author (that would be me!).
    (b) It will be easier to tell you from my own experience what I think is relevant, but of course, it is backed in great part by careful research, for which there is simply not enough space in this book to expand (see Rule 1).[4]
    (c) You will find other personal stories besides mine as case studies. These will help me illustrate how my "experience" has parallels with other innovators and increase the chances that you, reader, feel inspired by some or all of them!

---

[3] I hope my peer scientists will be lenient with some parts where I cannot provide the rigor of traditional scientific communications. I may oversimplify some concepts, but will always provide the source of inspiration or point to where the reader can find more information.

[4] You will find the most important references to help you do practical things.

2.3 Conceptualizing terms

2.3.1 Engineer, E_3

Let us start by establishing my definition of what an engineer should be now, or at least in the future. Do not bother at this point about the subscript "3"; it will be explained in Section 5.2.

An engineer:
1. is an active *connector* between science, technology and society;
2. matches knowledge and *ingenuity* to practical challenges; and
3. never stops *learning* and *teaching* others.

This definition is based on my interaction with other scientists and inventors who might or might not have obtained an "official" engineering degree. This means that a physicist or a dentist can fit in my description, as long as they are trying to solve problems, translating science to practice. But I also believe that you can be an "engineer" even if you did not have the opportunity – some would say privilege – of having obtained (yet) a traditional university education. In the end, different universities have such disparate study programs that it is hard to objectively compare all of them. I am aware that there are more serious and official or legally binding aspects to grant an engineering title, which I superficially cover in Section 4.2.

As an undergraduate student, or even before my university years, I was not aware of the differences between several professions and within science or technology fields. When I started traveling abroad, I was shocked to find that in some countries there were (unwritten) barriers built between engineers and scientists. The reasons of this unfortunate discrimination are still unclear.

> "Most engineering programs focus on standard 'engineering science' courses, such as statics, thermodynamics and circuits, that trace their influences back to the technological race with the former Soviet Union during the Cold War, as Jon Leydens and Juan Lucena explain in their book 'Engineering for Justice.' It was then – some seven decades ago – that engineering curriculums began to emphasize the scientific and mathematical basis of engineering, cutting back on hands-on engineering design and humanities courses. While most engineering programs now incorporate these types of courses, engineering classes themselves still often have a persistent divide between the social and technical." Future engineers need to understand their work's human impact [27].

I asked my PhD supervisor and lifelong mentor, Han Gardeniers, what he thought about this almost futile division between science and technology, or between applied or fundamental scientists and engineers:

I might be mistaken, but many of the great scientists, especially in the physical sciences, were also involved in engineering work. See J. W. Gibbs, who laid ground for our current understanding of thermodynamics, of whom Wikipedia says:

> "In 1863, Gibbs received the first Doctorate of Philosophy (Ph. D.) in engineering granted in the US, for a thesis entitled 'On the Form of the Teeth of Wheels in Spur Gearing", in which he used geometrical techniques to investigate the optimum design for gears."

So, if engineering and scientific skills are joined in one brain, why should we make the distinction? I think that the step from "how does nature work" to "how can we set nature to work for us" is not such a big one. It is probably inherent to the human thinking in his effort to survive the conditions of the planet.

Next to that, there is another famous quote from Theodore von Kármán (1881–1963):

> 'Scientists study the world as it is, engineers create the world that never has been'
>
> [28]

The key ingredients to success stories in human's history, in my opinion, are dialogue, collaboration and shared responsibilities. We will be covering all these aspects in this book.

Science, hoping not to offend any reader, is arguably the greatest achievement of humans. Like engineering, science is the result of modest steps at a time and the possibility to objectively communicate findings to other people within our lifespan, and more often after death.[5] Beware, I might be biased as an engineer in my fascination about science. I also believe humbly[6] that if you are reading this book, there is a high chance you will be part of future developments in science, technology and the advancement of society.

Science, technology and society *are definitely entwined*. For example, technology is supposed to come first, because humans were able to build machines without understanding the operation principles, by virtue of trial and error [104]:
- the start of the scientific revolution in the sixteenth century coincides with the development of the telescope and the microscope;
- a proper understanding of electricity was gained only after it was possible to manufacture machines able to produce it, and later to store it.

We can agree that society is then the sum of all human-related activity, encompassing the economical, juridical, the arts, and other dimensions – including science and technology.

---

5 By written texts, equipment built, or any other recorded form. I also say "objectively" because other forms of communication, such as in the arts, e. g., music, paintings or poems, are also universal, but more subjective, meaning that the message depends on your own interpretation.
6 If you find the topic of humility interesting, please check Sections 4.3 and 5.7.1.

2.3.2 Entrepreneur, E_2

Do not bother at this point about the subscript "2"; it will be explained in Section 5.2.

 Entrepreneurs are risk takers. The potential losses include the time of the entrepreneur and the resources devoted to change a product, method, practice or way of thinking.

An entrepreneur invests time in finding solutions, and risks to be ridiculed – by nature and peers – if the solution does not arrive in a decent amount of time, depending on the field. For example, a scientist's hypothesis may never be recognized or tested, or an invention may never become a commercial success. Entrepreneurs tend to have a vision and the dedication to stay on course, but typically can pivot when reality makes it difficult to reach their goals. For more on my take on success and failure see Section 5.11.

You can find in the literature many ways to define what entrepreneurs are or what they do. One I like in particular is that entrepreneurs turn technology and ideas into innovations in the market [130]. This implies that entrepreneurs create opportunities that lead to economic dynamism, economic growth and societal well-being.

I also want you to think beyond the classical gain for the risk takers who "exploit market opportunities"[7] and "whose function it is to carry out new combinations" [159]. There are many examples of important drugs that never became a commercial success, but the impact on saving lives is a much better measure to focus on. Please, also include nonprofit-oriented goals, under the heading of "social innovation," and see entrepreneurs as individuals taking a chance or risk in between people or processes in a broader sense (see Cases in Section 6.2).

There is a less romantic but powerful vernacular definition, which can be exemplified by a person who just goes out to sell food on the streets. This entrepreneur contributes to dynamic growth and well-being by keeping hunger at bay. In this book we will focus more on STEM-related types of innovations and entrepreneurs.

"*What is innovation?*," you may be wondering. Perez-Breva [148] says that:

> "defining innovation is a distraction from innovating (...) innovation: novelty with impact. But this is the 'definition' of what you end up with. You evolve toward it; it is not a given, and for that matter it is also a rather bad indicator with which to measure progress. At some point down the road, you'll bring something new to a community and its members will benefit from it. Their benefit will be your impact."

 There are, as you can imagine, many definitions of innovation. Simply put, it is when changes are made to something that already exists: a method, idea or product.

---

[7] In several occasions I have used in my lectures the example of C. Columbus, who had to convince the King and Queen of Spain to finance his mission to sail across the Atlantic Ocean in the fifteen century – after being rejected by the King and Queen of Portugal!

2.3.3 Knowledge, K

> Knowledge is information connected to a specific problem and *aligned* towards a possible solution. It may state a *what*, *why*, *when* and/or *how* of something related to the problem, and also *who* and *where*.

We must distinguish between information and knowledge, because finding information is arguably easier than acquiring knowledge. You may think of gathering information as reading material from a book or from internet searches, asking experts or your own personal experiences (see Figure 2.2). Formally, it has been defined that for information to become knowledge that relates to a problem, you must apply cognitive processing [121]. A small warning: common sense does not always help understanding how the world works. This implies that knowledge is different from common sense. This also brings me to a well-known pyramidal representation of the relationship between data, information, knowledge and wisdom (DIKW) [21].

What is important for you at this stage is to appreciate that different types of information allow building knowledge – often slowly – about almost anything. If you are disciplined and follow good methods while acquiring knowledge, you may become an expert in any field of your choice (see Section 5.6). The steps you follow will allow synthesizing or developing new knowledge. Knowledge is created as a bundle of heterogeneous but complimentary resources with three core types: explicit (documented information), implicit (applied information) and tacit (understood information). This interesting topic is still debated and redefined [56].

Moreover, the knowledge you acquire should help you transfer it to any different context or, if needed, enrich it with additional knowledge to solve problems in the future. One example is knowledge regarding needs and knowledge regarding solutions. Another significant element is related to practicing expert-like decision making, to develop a mental framework in the discipline of choice. These are summarised in the Pillars of Deliberate Practice Thrive, Ericsson et al. [82, 135] or as put by Ericsson and Pool

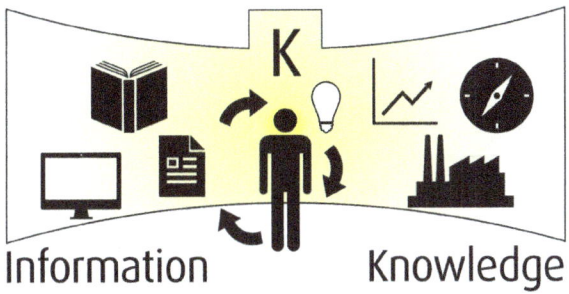

Figure 2.2: Information aligned by the "inventor" towards a solution becomes useful knowledge.

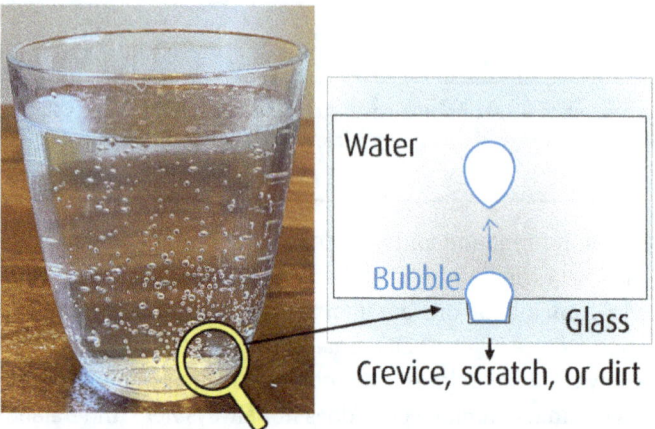

Figure 2.3: Gas bubbles can be trapped inside crevices, i. e., tiny cracks on the glass surface or particles covered with water. The author duly drank the water after taking the picture, to remain hydrated.

[81]: expertise can be learned with challenging but doable tasks, practicing elements of expertise with feedback and reflection slides. Thus, instructors should develop activities aimed at demanding substantial focus and effort from the students, instead of passive listening to a lecture or performing easy tasks (Wieman [182]). If you want to know more about these ideas, see [96] and Section 4.4.2.

? *Knowledge assignment:*
If you pour carbonated water (with lots of dissolved gas) in a dry empty glass, almost instantly you can see a stream of bubbles rising up to the upper surface open to the atmosphere. If you look carefully, there are points inside the glass surface that seem to be the source of these trains of bubbles (see Figure 2.3).

If you take a look at a single spot where bubbles emerge with a magnifying glass or microscope, you would most likely see that a small defect or tiny crevice – and sometimes dirt, *uggh!* – serves as the anchoring point for the sequence of bubbles to grow and detach, one after the other.

If you were tasked with getting more products from a reactor based on the presence of bubbles, using less electricity while controlling the way bubbles grow and collapse, *what would you do?*

The answer can be found in Appendix A, Section A.1.

2.3.4 Persuasiveness, P

 Persuasiveness is the ability to transfer understanding of an idea, which leads to an agreement between the persuader and the person or group of persons being persuaded. This is a key ability that entrepreneurial and innovating engineers need for developing relationships between people. I see two levels of persuasiveness, one where the persuader manages to make the other person understand something, and another where the person is *moved* to do something based on reasoning.

You may have probably heard about the importance of "soft skills." These have more recently been termed "durable skills," since they should last a lifetime [115, 116]. For more elements on this durable aspect, see Section 4.4.3. In contrast, some areas of technical knowledge learned during BSc or MSc degree courses will certainly have a short shelf-life because technologies evolve continuously, and educational programs cannot adapt fast enough. For example, programming courses purely based on a given programming language are obsolete quite fast. Logically, if your teachers and the course are good, it will not all be lost, because it can train you on structuring your ideas and enhance your logical thinking. Think of, say, a negotiation course based on human psychology that will probably teach you stuff with durability.[8]

In STEM curricula, it is common to have emphasis on mastering technical ability. Personal and interpersonal skills such as communication, teamwork, time management, emotional intelligence, and decision making are supposed to develop as the students complete their courses. However, in business schools, for example, it is easy to find courses explicitly providing knowledge about company creation, market forces, communication and negotiations [84]. In rare ocasions where durable skills are explicitly taught, we find *persuasiveness*.

Persuasiveness is instrumental to solve most societally relevant problems, which tend to be complex and demand multidisciplinary teams to co-create knowledge [155]. For example, when funding a research or business idea, persuading potential investors or governmental subsidy program committees is needed. To do so effectively, think of enlisting the assistance of experts or colleagues to sharpen your skills at persuading by reviewing your proposal, developing a perfect pitch and recruiting team members. You can find examples and tips in Chapter 5.

The innovator-entrepreneur must *convince the capitalist* that the higher revenues and/or lower costs stemming from his innovation will enable him to pay both principal and interest on the loan. The innovator must *convince himself* that the profits expected from the innovation will be sufficient to do this and leave a net profit for him. Thus, credit creation is a necessary but not by itself sufficient condition for economic development in a capitalist economy.

Joseph A. Schumpeter [158]

Compared to knowledge, this ability typically requires more time to acquire because it involves more than interacting with passive information. Persuasiveness depends also on the person's outgoing or social nature. It can be related to experiences since you must acquire the information, process it, act upon it and then reflect and adjust

---

8 This corresponds to my observation that human behavior (not sure about the brain's physiology) changes slower than the speed at which technology has evolved. For example, the stories we find in novels or books written centuries ago have not changed much. We keep doing the same, such as falling in love and going to war, albeit with shinier gadgets.

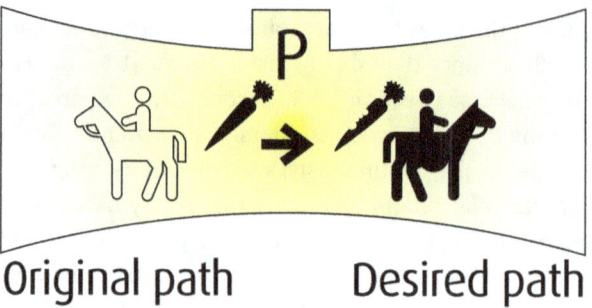

Original path   Desired path

Figure 2.4: With the right motivations or reasons, "persuaders" can change the course of ongoing processes.

future actions according to what did or did not work (see more in Section 5.2). For example, the value of persuasiveness can be quantified in terms of the number and quality of collaborations between research groups, the number of products sold or profit made by a salesperson, your ability to steer the direction of ongoing processes and the number of people persuaded and/or how fast the persuasion happened (see Figure 2.4).

Throughout this book, the term "persuasiveness" will be used mostly in the *positive context* of convincing others through reasoning that a problem one is trying to solve is important, or that a given solution is the right approach to solve it. Please be aware that persuasion can be found sometimes in negative contexts such as manipulation and propaganda. I discuss more elements of this in Sections 4.4.3 and 5.3.1.

 Persuasiveness assignment:
You are preparing yourself to enter a room full with highly learned jury members. The committee is composed mainly of scientists more mature than you, and you are not sure how inclined they are about the most commercially appealing aspects of your project. This jury will decide if you should get a handsome sum of money to execute your most desired project.

Your idea is based on a technology that can be used for many different applications, some more commercially rewarding than others. The scientific challenge is roughly the same, and as an inventor, you would not care too much at this early stage what the ultimate application would be – it is simply too far into the future.

You then have to decide how to prepare your deck of slides and which angle to choose while preparing your five-minute pitch. Which elements would you use, and in which order?
- ☐ The most commercially appealing (though controversial to some)
- ☐ Use a joke to set the tone
- ☐ Why you are the best candidate to get the money
- ☐ Thanking for the opportunity and the biggest scientific challenge

The answer can be found in Appendix A, Section A.2.

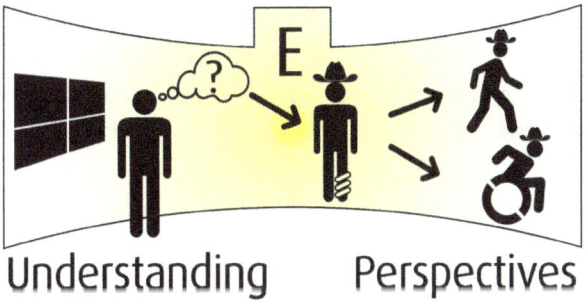

Figure 2.5: In this book, by empathy we refer to a person's ability to understand, imagine or even predict the perspectives or wishes of another person.

2.3.5 Empathy, E

Empathy is the ability to understand or accurately predict the perspectives of others. From that understanding, we can identify their needs, then decide to act on them. I actually propose in this book to call it *actionable* empathy.

There are different forms of empathy reported in the literature [109, 119], of which I selected the following:

Cognitive empathy: how well an individual can perceive and understand the emotions of another. This is also known as empathic accuracy. It involves "having more complete and accurate *knowledge* about the contents of another person's mind, including how the person feels," Hodges and Myers [109] say. "Cognitive empathy is more like a skill: Humans learn to recognize and understand others' emotional state as a way to process emotions and behaviour."

Affective empathy: expressing an understanding of how and why the other person feels in a certain way.

Emotional empathy: this consists of three separate components [65]:
1. feeling the same emotion as another person;
2. personal distress, referring to one's own feelings of distress in response to perceiving another's situation; and
3. feeling compassion for another person; this is the one most frequently associated with the study of empathy in psychology.

Behavioral empathy: a demonstration of active listening, expressing the desire to understand more about the feelings, experiences or reactions of other people.

Contagion empathy: characterized by relative passivity to the emotions of others.

Projection empathy: characterized by active engagement with the emotions of another.

I tend to agree with the projection empathy, which is very close to my actionable empathy, because it is capable of producing greater understanding about other people or the impact of the problem. Based on a recent review [119] of the book "Being Me Being You: Adam Smith and Empathy" [100], I understood that a combination of projection and contagion is desirable. Contagion empathy provides raw materials needed in imaginative and projective exercises and compensates the difficulty to have direct access to the emotions of others.

> "Durable skills" are gaining a critical role in the STEM field, with the term "STEMpathy" being popularized by Friedman [101]. Friedman argues that technology- and knowledge-based capacity in STEM must be coupled with empathy for other human beings, since "nobody cares what you know, because the Google machine knows everything"; instead, Friedman stresses that the future is about how we apply what we know.

STEM professionals who can empathize or have the ability to place themselves into an imaginary situation and somehow feel those emotions, or imagine them at least, should be able to better understand the context of a problem. Pursuing an empathic approach can help to develop the ability of the engineer to formulate each problem succinctly and effectively, assisting in understanding the "why" of the need to solve the problem. I also believe we should be able to expand our empathy beyond other fellow humans – currently living and future generations – and consider also other living organisms.

> "There are still gadgets commercialized today that were badly designed, almost forcing the user to read boring manuals before using the gadget. For example, the user-interface of an electronic appliance may not be intuitive, arguably a result of designers not having enough empathy for the end-user."
>
> Jim Heirbaut, journalist for De Ingenieur (https://www.deingenieur.nl), a Dutch magazine for engineers
>
> "Economists have made the problem of only looking at the Wealth of Nations written by Adam Smith [162]. Luckily and increasingly of late, we too understand the role of empathy and sympathy based on this masterpiece theory of human behaviour laid out on Theory of Moral Sentiments [161]. Just now we return to the ideas Smith proposed after decades of going into the wrong direction: cost-benefit and numbers only."
>
> Robert Lepenies in private communications, January 2021
>
> "I think empathy is not necessary for problems that are being noticed by society or individuals, and as such are available to (entrepreneurial) action. However, what about the problems we don't see? It could be argued that venturing out (empathetically)[9] and being able understand the problems opens up many more opportunities rather than the statistical chance of having firsthand experience, and as such *Knowledge* being enough.

---

9 Empathetic vs sympathetic vs empathic (https://www.grammarly.com/blog/empathetic/). According to Grammarly, the words empathetic and empathic mean the same thing. Empathic is a slightly older word, used first in 1909; empathetic is from 1932.

An example of this might be Slumdog Millionaire (https://en.wikipedia.org/wiki/Slumdog_Millionaire), where the quiz contestant from the slums by sheer luck knows all the answers since he experienced them in one way or another himself. However, how likely is that? So I would argue Knowledge is not a must, however it is a multiplier effect that increases the amount of opportunities discovered. The problems that are already known in your backyard do not require high levels of empathy to see. Instead, it can really help looking over into the neighbours' yard."

Bas Koelewijn BSc, focuses on inclusion and innovation

Not surprisingly, definitions of sympathy occasionally overlap with those of emotional empathy. Smith actually used the word "sympathy" for what later was labeled "empathy" [119]. We can define *sympathy* as a temporary emotional response coupled with an expression of that emotion toward the other individual, for example, pitying an individual after the loss of a loved one and offering condolences.

So if you meet me
Have some courtesy
Have some sympathy, and some taste
Use all your well-learned politeness
Or I'll lay your soul to waste, mm yeah

♪ ♩ Rolling Stones (English rock band) – Sympathy for the devil, 1968.

With sympathy, the emotions felt and expressed do not necessarily reflect the emotional state of the person who is experiencing a reaction. In the context of the Rolling Stones' lyrics, you may realize there is a potential advantage for you to be sympathetic. After all, you may want to avoid trouble with the devil, but you may not feel that inclined to empathize with such an evil spirit. If you want to read more about this and other cultural references, check Sections 4.4.4 and 5.3.

A way to gain awareness over empathy:
I invite you to think about a situation in which you were confronted with a problem and you had to be emphatic.
   Now, you can actually do an assignment I use in my courses:
Empathy assignment:
Please write a document of about one page or less (including figures) describing a situation faced by a real or hypothetical individual (member of public, patient, customer, etc.), and on a second page, how a solution of your choice changed (in reality or hypothetically) their situation. Wherever possible or needed, you may use references from the scientific literature or reliable media sources or websites to back up your statements. If it is a personal experience, this may not be required.
   The main aspects you need to pay attention to when preparing your written document are:
1. the quality of websites/sources consulted;

2. the way you perform the storytelling of both scenarios (it should be authentic; see examples and tips further down);
3. make sure you use images or schemes to illustrate the problem and solution and that they are clear and of good quality.

You also have freedom to define your own problem and solution, but here are some pointers:
1. Define a problem of a person or company.
2. Define a solution for the stakeholder(s) defined in point 1.
    - Think of the pros and cons of the solution for the stakeholders.
    - Would you do it like you already imagined or thought of in point 2, or would you like to adjust something?
3. Think of the reaction of the stakeholder.

However, if this strains too much your neurons, here you can find some options as inspiration.
- A villager lives close to a chemical plant with a refinery or other specific nasty reactor. This person suffers from a serious health condition due to the contamination of the air or water in his surroundings.
    - You propose a change (reactor, process, etc.) or intervention, and this leads to an improved quality of life of the villager. You then describe the situation from the perspective of the "villager."
- A patient suffers from a type of treatment that is very important for his quality of life.
    - You want to propose an innovative product that can change this situation.

Where to look for information and make your case:
- You can use your own network of people to ask for personal experiences.
- You can look at reliable news sources (interviews, documentaries).
- To propose the solution, you can tap into the scientific literature, perform a patent search or talk with/write to companies working in that particular field.
- Your teacher, of course.

Answers can be found in Section A.3.

2.4 Mixing the ingredients: K+P+E

The definitions of engineer, entrepreneur, knowledge, persuasiveness and empathy that I just shared with you allow me to combine them in a comfortable way. The ability and possibility to learn from inventors or scientists from the past also allowed me (and certainly many others) to imitate actions or assimilate concepts that I found relevant.

Similarly to artists, architects and other professionals, we learn a lot by imitating. For example, biomimicry is a great example where inventors copy from Nature in the search for solutions to technological problems faced by humans. The beauty of technology and scientific development is that sometimes it is even possible to keep some basic concepts, but change the actual objects of study, and by making analogies and through hard work, we gain insight into new phenomena.

2.4 Mixing the ingredients: K+P+E — 17

A very simple example where the KPE can be seen and mixed.
K: You come up with a (brilliant) idea or solution for a problem.
P: You have the opportunity to pitch your plan or idea at a Dragons' Den show.
E: You prepare your pitch in such a way that you could persuade a scientist or a commercial/investor expert.
   You would need to anticipate different scenarios, make sure you connect with ongoing social, technological and scientific developments and bring it all together.

Let me tell you about what I interpret as innovation and similarities in research. A study I performed with collaborators on water jets impacting liquid droplets resembles the work of Harold "Doc" Edgerton on high-speed photos of a bullet fired through an apple (https://news.mit.edu/2021/water-jet-droplet-injections-0818).

Edgerton produced beautiful and explosive detailed images with a technology he developed based on flash lights. This technological feat enabled him to capture sequential images of a bullet being shot through several objects, including bananas and several apples. In Figure 2.6 you can see two engineers in a similar pose, in front of an experiment, but for completely different purposes, decades apart – 62 years to be more precise.

Our new videos, of a water jet fired through a droplet, revealed similar impact dynamics and gave us better insight into the physics at play (see this video [https://youtu.be/dRCaJ_kWKkQ]). Since the droplets in our experiments were transparent, we were able to track what happens inside a droplet as a jet is fired through [149]. This analysis has helped us tune needle-free injection systems, the foundation of a company, that is further expanded in Section 3.4.2. Here I am highlighting the entrepreneur character defined above, which goes in hand with the ambition to solve problems and bring them to society.

Allow me then to draw three lines that connect my proposition of engineering with the KPE definitions (see Figure 2.7).
- The link between "*connector*" and "*empathy*" is almost self-evident, but let me expand a bit. Empathy (as defined previously) allows the engineer to identify the needs of a stakeholder or a section of society, and connect technical advances with useful applications to solve the stakeholder's[10] or societal problem.
- Ingenuity, which corresponds to the quality of being clever, original and inventive, initially made me think of the Spanish translation of the word "naive," i. e.,

---

10 What I mean with *stakeholder* and the importance of aligning interests and manage expectations is: "Academia is often focused on pushing the boundaries forward through proof-of-concept demonstrations, while corporates may be looking for something incremental that can align to existing offerings, mature development processes or sales motions. And in between are start-ups coming from universities that are in the process of establishing product-market fit." – Ram Jambunathan, Managing Director, SAP.iO in How to build an entrepreneurial university [29].

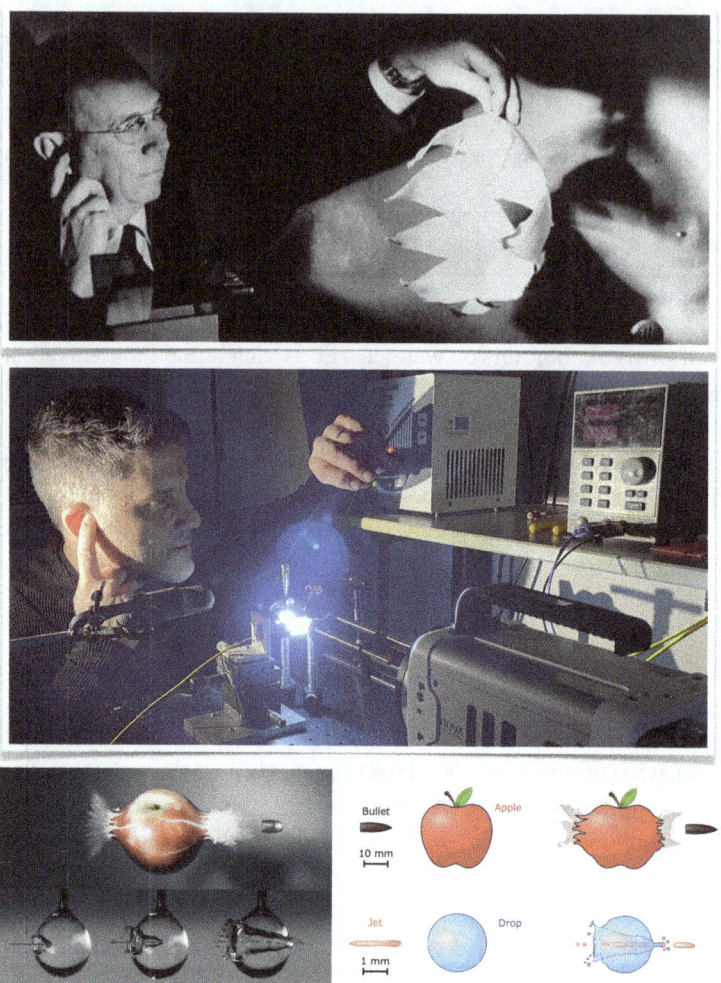

Figure 2.6: (a) Self-Portrait with Balloon and Bullet – Harold E. Edgerton, MIT, USA, 1959: Harold E. Edgerton creates a whimsical self-portrait protecting his ear as a bullet pierces a balloon, with an ultra-high-speed flash at an exposure of one microsecond (one millionth of a second). "Stopping Time" (1987), p. 33. from Harold E. Edgerton Collection. Courtesy MIT Museum. (b) Portrait with Drop and Jet Twente, The Netherlands, 2021. David imitates the pose of Harold E. Edgerton with a new setup to generate microjets for needle-free injections (https://news.mit.edu/2021/water-jet-droplet-injections-0818) [149]. The droplet and jet are too small and too fast for his mobile phone camera to capture it. See left inset. The most apparent difference between both experiments is the fact that we did not have to clean the residues of exploded apples in our lab.

An Engineer …
is an active **connector** between Science, Technology and Society, matches knowledge and **ingenuity** to practical challenges, and never stops learning and **teaching** others

Figure 2.7: Alignment between the engineer definition from Section 2.3.1 and the KPE framework. Please note that the colors used to highlight each word are not implying unique or direct analogies between the definition of an engineer and the KPE. Adapted from [96], Creative Commons CC-BY license.

ingenuo. To me, a somewhat naive approach can help finding information and transforming it into knowledge that can be used to find solutions where serious or established approaches might not work. One example of a seemingly silly approach is the carbonated water assignment on page 10, where for decades of scientific and technological progress in ultrasonic sonochemistry (i.e., chemistry caused by sound phenomena), that idea was not explored until a few years ago, as will be shown in Section 3.4.

– A good teacher can persuade those at the receiving end to improve their understanding and take action towards the solution of a problem. Persuasiveness can be seen as a positive interpersonal trait, whereas ingenuity is more of a personal trait. Having the right ingenuity can assist teaching facts and ways of working to other people and yourself when in a reflection phase (see Section 5.6).

Lastly, a subtlety in my definition that may be lost for many readers comes from the fact that the word for "teaching" is equal to the word for "learning" in Dutch and Old English.[11] Those who have had the opportunity to teach will agree with me that we learn a lot while we teach others. This is where I think we can have a positive impact on the future of engineering. Engineering classes are typically focused on decontextualized problems, and we need to always try to give a social context (see more in Chapter 4).

Looking back in time, we can see how the first industrial revolution was about creating machines, which were used to help mechanized manufacturing, providing energy through steam.

After having the first batch of technology or machines, engineers improved them over the years. Until now, we could argue that having the right knowledge was crucial. But today we know that we need to think more about the impact of our technological and societal development on the environment and society.

---

11 It can apparently be traced back to Proto-Germanic, or even Proto-Indo European [30].

Therefore an "engineer of the future" needs to think about, and include, other perspectives besides technical skills in research and design activities. For example, we have to think more about the environment, but why is this important in your view?

The answers to this "rhetoric" question given by some of my collaborators can be found in Section A.8.

2.5 KPE against epidemics

Here is an example of serious consequences when a good balance of KPE is not reached. You will have to excuse my extended focus on this particular "ancient case" because of three reasons:

1. it is highly relevant to demonstrate the power of teamwork in (medical) science, particularly when a problem affects the lives of millions of people;
2. it concerns a case of misrepresented facts affecting a scientist in political disadvantage;[12] and
3. a virus is at the center of attention, and our world continues to face risks related to tackling diseases that affect us at a global scale.

Facts of an underdog story

Figure 2.8: Carlos Juan Finlay de Barres. Public domain, English Wikipedia (https://commons.wikimedia.org/wiki/File:Finlay_Carlos_1833-1915.jpg). This file is licensed under the Creative Commons Attribution-Share Alike 4.0 International license.

Carlos Juan Finlay de Barres (1833–1915) was a Cuban epidemiologist doctor who studied in the USA and France, in a period when Cuba was still a Spanish colony (Figure 2.8). Finlay is known in the scientific world as the first to explain how yellow fever spreads: via mosquito bites. However, even though he did careful research and demonstrated the validity of his ideas, his work was largely ignored for many years.

---

[12] The fact that I was born in Cuba explains why I want to highlight this less-known story about the discriminations imposed by the dominating colonial powers Spain and United States during the nineteenth and twentieth centuries.

Finlay actively collaborated with the first research commission from the USA to investigate the causes of yellow fever in 1879. On 18 February 1881, he presented in Washington a theory that antagonized with the most accepted possibilities: direct contagiousness from being in contact with a sick person and the anti-contagiousness theory attributing it to environmental conditions, or "miasma." He postulated the existence of an agent, independent of the disease and the infected people, capable of transmitting the disease. On 14 August 1881 he presented to the Royal Academy in Havana, and later published the hypothesis that a mosquito would be such transmission agent, narrowing the search to the female of *Aëdes aegypti*.

However, this theory was not tested with experiments for 20 years! This happened because of a combination of the novelty of Finlay's epidemiological proposal with the fact that the scientific community was focused on finding the microorganism causing it, presumably a bacteria. Only years later, in the late 1920s, it was found that it is a virus.

Finlay and his collaborator, the Spanish doctor Claudio Delgado Amestoy, did experiments between 1881 and 1900 to verify the "mosquito idea." In parallel he published internationally the main measures to prevent epidemics[13] caused by yellow fever, aimed at destroying the mosquito larvae.

The USA government sent two more commissions to Cuba to investigate the yellow fever outbreaks in 1889 and 1899, which largely ignored Finlay's work. However, by the end of 1899, George Sternberg, the General Army Surgeon of the USA, created a new commission during the military occupation of Cuba. Sternberg knew Finlay since 1879, but did not agree with his theories, and the "mosquito theory" was not included in this fourth Commission. In July 1900, two British doctors visited Havana and suggested verbally to the commission to pay more attention to Finlay's theory; they published their opinion in England two months later.

The Commission visited Finlay's home in August 1900, who shared his publications, gave recommendations and donated mosquito eggs from his home lab. Members of this Commission took up the risky decision of infecting themselves using the mosquitos – apparently without proper authorization. Jesse Lazear died 13 days later, having carefully documented in his diary the first experimental demonstration of the "mosquito theory."

Suddenly, Walter Reed, the president of the Commission who was also skeptical until that moment, seized Lazear's diary and presented a "Preliminary note" on 22 October 1900 during a scientific event in the USA. This note, which presented Finlay's theory as valid, was not produced with scientific rigor. Reed claimed that Finlay did not demonstrate it, even when he had reported a similar case in 1881, because the

---

[13] An epidemic is the widespread occurrence of an infectious disease in a community at a particular time, whereas pandemic refers to a whole country or the world.

incubation period of Finlay's study did not match the evidence, effectively discarding 20 years of work.

In 1932, an independent study demonstrated the validity of Finlay's results, which varied due to a temperature dependence of the incubation time. Reed used Finlay's theory and suspected mosquito's in other rigorous scientific studies, up to the point that in correspondence letters he started calling it "his theory." Reed was publicly recognized in the USA as the discoverer of the cause of yellow fever. He died in 1902, before he could demonstrate that *A. aegypti* was "the only" possible infection agent.

Such evidence came surprisingly from the epidemiological measures proposed by Finlay and enacted by the American military doctor William Gorgas, who rightly gave credit to the Cuban doctor for helping eradicating yellow fever from Havana in 1901 and later from Panama.

When the American army occupation ended and Cuba became independent, Finlay took charge of the Health Department. Within just three months, in 1905, he succeeded in stopping the last yellow fever epidemic ever recorded in Havana. The last case of yellow fever was reported in 1909.

Between 1905 and 1915, Finlay was officially proposed seven times for the Nobel Prize, which was not granted. He got, nevertheless, many other honors and recognitions from French and British institutions. This section was largely written based on [1, 53].

Underdog KPE analysis

There are details that we do not know today, more than a century after the events described in the previous section. But if we stick to the summary I provide about Finlay's ordeal making his voice heard, this is the analysis I give in the context of KPE.

Knowledge The suspicion Finlay had about the mosquito being the transmission agent was not unique in itself. There were other diseases and studies in the same period where it was known that mosquitos were disease vectors, e. g., malaria by the *Anopheles* mosquito. However, not all stakeholders (colonial Spain rulers, occupational representatives from the USA army and scientists) agreed on the validity of the knowledge each of the others had.

Persuasiveness I am not aware of what else Finlay could have done. He traveled to several conferences and published in Spanish and English, trying to persuade the scientific community and other authorities. Finlay's willingness to collaborate is evidenced also by sharing his mosquito egg samples, and always keeping an elegant posture: I have not found any reference of him complaining about what Reed did. You can see that the persuasiveness of the British visitors induced the visit that the Commission's members paid to Finlay, finally setting the experimental confirmation of an old theory.

Empathy Here I believe lies the missing ingredient. My interpretation is that the Spanish authorities – before losing Cuba as a colony to the occupational USA army – and the members of the Commissions looked down on Finlay.

It is not new that rulers and occupiers tend to treat the people from the territories under domination with less esteem than others. Why is it then that it took external experts to suggest testing empirically Finlay's old theory?

As could be expected, the colonial rulers and the Commission members had interests of their own, as reported in the literature, ignored Finlay's claims and marginalized the evidence he had. During the Spanish domination, it would have looked bad on the Spanish Crown to acknowledge many of the mistakes around the epidemics. Actually, their priority was the ongoing independence war fought in Cuba.

The Commission was also trying to validate theories of their own, such as identifying the actual source of the disease, which many thought was a bacteria. To make matters worse, even when the evidence was in Finlay's favor, Reed did not work together with the Cuban scientist. Instead, he wrongly appropriated the knowledge accumulated for 20 years by Finlay and compiled in the diary written by the unfortunate Lazear, who died experimenting on himself.

- The worst of it all is that so many lives were lost to this pandemic, and many could have been avoided with a bit more empathy and good teamwork.

Another, more recent, KPE analysis of a real-life situation not experienced by me or my network can be found in Section 5.8.

During the preparation of the example of Finlay and the yellow fever epidemic, my colleague Devaraj van der Meer raised a related case closer to our times:

> This case of Finlay is clearly an example of a problem from the past, resulting from insufficient esteem for scientists from the colonies. But this type of problem persists until today; for example, a bacteriologist or virologist will emphasize the role of the infecting agent as the cause of the disease, whereas a fluid dynamics scientist or doctor may emphasize the means of transmission. It is simply that humans tend to focus more on the aspect that they know a lot about, and tend to marginalize other aspects.
>
> Things get dangerous if an expert is close to decision makers such as a government during a pandemic and that person is heard best – or has the loudest voice. It is for this reason that many countries, including the Netherlands, made suboptimal judgments about wearing masks during the COVID-19 pandemic. While many countries around the world were clearer about the need to use masks at all times, the Dutch government listened to a virologist with limited appreciation for the way in which viral diseases are transmitted, and a complicated set of rules led to confusion among the population as to when and where the mask should be worn.

I agree with Devaraj, and I think that a clearer mandate to wear masks could have saved many lives, not only by reducing the transmission of COVID-19, but also limiting the impact of other infectious diseases – seasonal flu or colds – at a time when the healthcare sector was overloaded.

I just wonder how much more fluid dynamics experts would have to do to make their voice heard. There are plenty of scientific publications and news items about this, but if you want to know more, there is a link to a special webinar organized by the Royal Dutch Academy of Arts and Sciences (KNAW) on June 4, 2020. The videos (https://knaw.nl/nl/actueel/terugkijken-bijeenkomsten/corona-van-druppeltjes-tot-pandemie) cover questions such as "What is the role of droplets and aerosols in spreading the corona virus?" and the experts discuss the transfer mechanisms and remedies.

2.6 Applicability of ideas

Based on traditional education practices, a common starting point when engaging in problem solving is knowledge. However, it is not necessary to be an expert at the problem you are trying to solve from the beginning. Instead, you learn during the process, and you therefore can start with just a piece of knowledge, after being persuaded by someone else or inspired by an empathic painful problem. Arguably, acknowledging your knowledge of the subject is "incomplete" may prevent so-called tunnel vision, where people focus mostly on what they have seen before, and allows for devising or designing new solutions to a problem.

Furthermore, the three ingredients of KPE are not linear or static concepts as might be inferred from the sections above; they are entwined. For example, the ability to persuade is often predicated on connecting with those being persuaded; namely, empathizing with them. As we empathize, we increase the chance of gaining more knowledge first-hand about the problem in question. This can then become an iterative process, hopefully in a virtuous cycle.

Figure 2.9 provides a refined visualization of the interaction of the ingredients in the KPE framework. On the z-axis (in plane), different disciplines or fields can be assigned, because the field of innovation has to deal with multidisciplinary steps, as we will explain in Section 3.1.

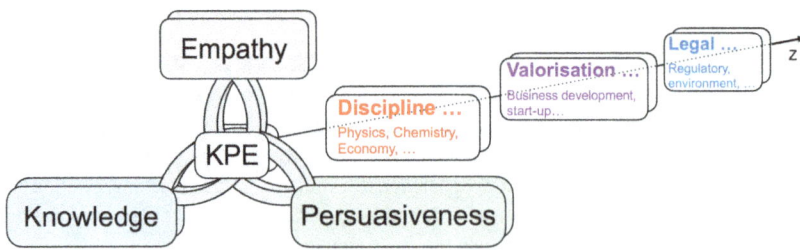

Figure 2.9: Schematic representation of the interdependence of knowledge, persuasiveness and empathy. The importance of each ingredient runs across all fields of activities, particularly for innovation (in plane – z-axis). The order of the fields is purely arbitrary in this figure, but in some cases the legal aspects need to be covered before any valorization action is taken. Adapted from [96], Creative Commons CC-BY license.

Any real-world situation demands constant back and forth from all areas (z-axis) as time passes. It strongly depends on which was your starting point: valuable knowledge, having been persuaded by a start-up founder or perhaps after empathizing with a stakeholder suffering from a problem.

The interdependence of K, P and E seen in Figure 2.9 is the foundation to solve problems and to teach how to create innovative solutions, S. There are a myriad of other factors (OF) which can impact the applicability (A) of a given solution at a given time (t). Therefore,

$$A(t) = S\big(K(t), P(t), E(t)\big) - OF(t), \tag{2.1}$$

where there can be a time dependence on each of the ingredients as well. A short analysis of how each ingredient's importance may vary over time is given in Section 2.9. Specifically, the importance of each ingredient relative to each other can vary depending upon what stage of the innovation process one is at. Indeed, once we identify OF and quantify them or establish their qualitative relevance, you could say they are moved to the K part, increasing the Applicability. The importance of OF and ways to quantify them comes next in Section 2.7.

2.7 A decision making tool

When you can measure what you are speaking about and express it in numbers, you know something about it; but when you cannot measure it, when you cannot express it in numbers, your knowledge is of a meagre and unsatisfactory kind; it may be the beginning of knowledge, but you have scarcely in your thoughts advanced to the state of science, whatever the matter may be.

Lord Kelvin, British scientist (1824–1907). The quote has been published under Popular Lectures and Addresses vol. 1 (1889) "Electrical Units of Measurement," delivered 3 May 1883, Oxford [31]

This quote raised an interesting discussion with one of my undergrad teachers in Cuba and good friend Alfo J. Batista Leyva. I was not aware of the fact that the last few sentences could be interpreted as divisive and not what you would consider "empathic" to other fields of knowledge. I agree to that and would like you to critically think about the message transmitted.

There are branches of scientific research that cannot rely solely on quantifiable aspects, for example in history and psychology. I believe Lord Kelvin was probably thinking about a smaller subset of science and research activities, during a period where he was working on defining quantities, measuring properties for which there were maybe some instruments or new ones were being built.

I hope that after you have been provided with the basics of the KPE framework, it can guide your thinking and your approach to innovation or problem solving. There are numerous paths you can follow to practically apply this framework. In all cases, however, you will be have to make decisions. Decision making is a complex process and a very active field of research.

Often the interaction or comparison of factors influencing a decision is unclear. This is the case in innovation processes, where clearly defined impacts on outcomes from one factor versus another are often difficult to establish. Unfortunately, uncertainty is much higher in the beginning of an innovation process, when precisely many decisions have to be made, because changes are more costly if they occur later.

When deciding on what to innovate, especially in the early phases of a project or with less experienced teams, it is important to demonstrate that a new idea is better than existing alternatives. The easiest answer you may get in the business or industrial worlds, but also at a household or personal level, is cost: *the lower the better for a given quality level*. This is of course assuming that other important elements are kept alike.

Without entering the field of economics or summarizing all the theories and models that explain why some people choose one product over others, let us agree that cost is not always an important factor in the choices we make, particularly if safety is in the picture.

In the technical or production world, you might want to maximize throughput, the efficiency of a given reaction or safety for the people working and the environment. Sometimes, these drivers can be quantified easily, because there is a way to measure them that most people would agree upon, or an objective measurement which can be performed.

Logically, it is more difficult when more than one person is involved in the decision making process.[14] This happens because every stakeholder may have different motivations, or drivers (see Figure 2.10). It can be increasing efficiencies, reducing costs, mitigating ecological impact, reducing the number of workers in a costly or dangerous occupation, and so on.

You can compare the prices of two fruits at the market, or the increased purity when producing some commodity. But what about when there is no objective measurement to be made? Or the complexity or dimensionality of the measurements are such that they cannot be directly compared?

For example, what to do when we feel something is better or worse, or rank apples and pears if they have the same price at the market? How important is the feeling of safety or sustainability? Or some experiences associated with subjective perceptions, such as while driving a particular car model?

Sometimes we can only say that one fruit tastes fresher and felt softer in your mouth, or that it felt more comfortable or safer while sitting inside a car. But these assessments are highly subjective. The ultimate question is: how can we put together all these quantities or qualities under a unifying number? Answers to these questions are provided in this section. But before you continue reading, pause for a moment and reflect about these questions while looking at Figure 2.10.

---

14 Systems thinking supports the idea that complexity levels increase when more people are involved in decisions, a fact often overlooked by companies or research teams putting more people than needed on the job of finding solutions to complex problems. A proper alignment can be reached with FunKey architecting, where key drivers of the stakeholders and functions of the system are aligned in the early design stage [128].

Figure 2.10: How would you choose which fruit to buy if the price is the same and you will share it with more than one person?

I hope you agree with me when I say that it can be useful to have a simple-to-use method for helping to align the interests of different stakeholders, and increase transparency when different team members are taking decisions. The intensification factor (*IF*) method introduces a guide that is currently lacking in several fields: deciding when to continue going through an innovative process.

I have presented *IF* as a tool to assist in deciding when to intensify a process, for example, when to continue expending resources in order to create something new. It is one out of five steps proposed in prior work done with my collaborators [86, 95]. Feel free to replace the 'intensification' word with innovation if you like.

The simplicity of the method enables evaluation from minimal information (economic, technical, social, scientific, etc.) during the early phases of a project, and it can grow in levels of detail as progress is made. The strongest advantage in the context of this book is that it can be understood by outsiders and nonexperts without much effort.

The *IF* method is highly applicable when there is a changing dynamic of factors to be considered, for instance:
- *When information is initially limited* but grows over time, i. e., as knowledge and additional factors become known.
- *When presenting information to different audiences*. The full scope of all factors does not need to be given, but only factors which are significant to those being shown the information.
- *When there is uncertainty* about the information being accounted for, be it about the accuracy, about the significance of the information or if an objective quantification is not possible and therefore a factor's value may change depending on a subjective basis.

The *IF* method hinges on quantifying the decision process into a single number, where different aspects need to be factored. A factor F is any characteristic or aspect that is an input into the creation of an innovation or solution. This is intentionally broad so as to widen the applicability of the methodology. However, F should ideally be quantified with a value for a "before scenario" and an "after scenario," to provide a basis of comparison as to whether a meaningful improvement occurred. Let us represent these values as F_b (before) and F_a (after).

As I mentioned earlier, one common quantifiable F considered in many cases is cost. The simplest case for decision making is to only consider one factor, e. g., one could make a decision purely by considering the F_b cost and the F_a cost of a scenario or solution. However, decision making is often far more complex, with the desirability of an outcome or solution having subjective components. We attempt to capture the desirability, d, in this method, where for the sake of simplicity, $d = 1$ if a decrease in F is desired and $d = -1$ when an increase is desired. Therefore, the intensification or *innovation* factor is calculated as follows:

$$IF = \left(\frac{F_b}{F_a}\right)^d. \tag{2.2}$$

If $IF > 1$, an improvement has occurred or the innovation future scenario will result in a better result. The last step is to calculate all the known (n) categories or IF_i, e. g., cost, temperature and risk, and multiply them: $IF_{\text{total}} = \prod_{i=1}^{n=3} IF_i = F_{\text{cost}} \cdot F_{\text{temperature}} \cdot F_{\text{risk}}$. Since F_b and F_a of each corresponding factor will have the same units, the resulting IF will all be dimensionless.

! Several assumptions simplify the application of the *IF* method and allow us to obtain weird outcomes:
- each factor F can have an objectively attained numerical value;
- F is nonzero (dividing by zero is a no-go);
- F follows a linear scale (other options are beyond the objective of this method and book);
- the desirability exponent d is either 1 or −1 (other values can improve balance or put stress where it is most needed).

In the case where F does not have a numeric value or it is difficult to calculate an accurate value, a symbol can be assigned. For example, a factor viewed in a positive way or deemed to have a positive impact on an outcome or solution can be assigned an arbitrary number of "+" symbols. A factor which in comparison is deemed less desirable can be assigned fewer "+" symbols.

Assuming a linear relationship, these symbols can then be converted to values. Exactness or accuracy is not the point in this scenario. The purpose is to provide a basis for quantifying the relative change in value occurring from F_b to F_a. Alternatively, ranking several nonquantified F along a scale of values can also be done to assign numeric values for comparison. To avoid the possibility of manipulating the final result or limit endless discussions on how many "+" symbols to assign to a given factor, you could also split qualitative from quantitative factors.

Note that F can take on a value of zero in many scenarios, e. g., when a quantity crosses from negative to positive or vice versa. To avoid $F = 0$ in the denominator, rescaling, turning F into a percentage or changing units (e. g., degrees Celsius to

Kelvin) can often be performed. Finally, alternative values of *d* can be assigned depending upon the stakeholders involved in the decision making process. What I mean is that a group of experts can define and agree on changing *d* at any time. This can be interpreted as assigning weight functions, based on the importance of a given factor.

Please be aware that the final *IF* number is not a predictor of reality, but a means to facilitate clear discussion by multiple parties making a decision, where they can reach consensus on which factors to include and to exclude, how the assignment of values can occur, etc. The final decision to innovate or execute the new solution can then be taken by the team speaking the "same language" based on the factors agreed upon.

Instructions to build your IF calculation table.

Table 2.1 illustrates the steps and required values in order to obtain an individual impact factor *IF*.

Table 2.1: Suggested steps to calculate *IF*.

Factor	Before	After	Exponent	Fraction
F	F_b	F_a	d	$IF = \left(\frac{F_b}{F_a}\right)^d$

$$d(F) = \begin{cases} +1 & \text{if a decrease in } F \text{ factor is desired} \\ -1 & \text{if a decrease in } F \text{ factor is undesired} \end{cases} \quad (2.3)$$

The meaning of the absolute value of *d* has to be determined depending on the specific target or goal. Experts would have to agree on, e. g., if safety, cost or commercial considerations have a stronger relevance. If such information is not available or an agreement cannot be reached, it can be set to unity as we have assumed for all the example cases presented in this textbook.

The intensification factor for a given number of *n* changes can be calculated as follows:

$$IF = \prod_{i=1}^{n} \left(\frac{F_{b_i}}{F_{a_i}}\right)^{d_i}. \quad (2.4)$$

Independent IF_i values can be calculated for each change in consideration, for example, ecological impact and economic benefits of a given equipment or process under analysis for its improvement or replacement, which may include a longer channel, use of different materials, improved safety, etc. For each of these factors, you just need to add a new row, fill up the values you have measured or estimated, define the exponent value, calculate the independent factor and multiply them all.

The total *IF* of a global intensification initiative having a number *p* of potential intensification strategies can be calculated as

$$IF_{total} = \prod_{i=1}^{p} IF_i. \quad (2.5)$$

Lastly, a short movie with a more elaborated explanation can be found at https://youtu.be/B7ZyATa1-ng – password: EEE.

City trip case

Let us define a very simple example where a group of friends is going on vacation and needs to agree on the location to visit and the duration of the visit. The main aspects to consider are the total estimated budget for the students to spend, which is related to the means of transportation and accommodation, as well as the total duration of the trip and location (see Figure 2.11).

Figure 2.11: Three traveling students deciding where to go: Diana, Nicolás and Thijmen.

There are two city destinations, A and B. For B it is very easy to get a visa, but is more expensive and the trip is shorter. These will be the factors that will be put together to compare the value of each alternative (see Table 2.2). For example, ticket prices of different transportation means, the cost of a room, independent or shared, visa requirements, etc., can be included. They even want to have a trip as long as possible because the vacation time allows for it.

Table 2.2: Intensification factor calculation for a simple comparison.

Case	Factor	City A	City B	d	IF
Student trip	Ease of getting Visa	+	+++	1	0.33
	Total budget [euro]	300	150	1	2
	Trip duration [days]	10	3	−1	0.3
				IF_{total}	= 0.2

The fact that IF_{total} is smaller than "1" indicates that A is probably a better option than B, at least considering the factors listed, despite being cheaper and the easily obtained visa. This "situation" is further expanded in note 1, Section 3.1, page 41.

2.7.1 Origins of IF

"Should I stay or should I go now?
If I go there will be trouble
An' if I stay it will be double
So come on and let me know"

The Clash (English punk rock band) – Should I Stay or Should I Go, 1981.

I want to let you know a bit more about my motivation to study the topic on "how to decide." It goes beyond the "fruit selection" and "city trip" examples from Figures 2.10 and 2.11. It all started when I was preparing a new course in 2015. While reading a textbook to learn about a topic entirely new to me, process intensification, I stumbled upon a table with two columns and several rows, listing the different specifications or features of two chemical reactors [151] – please check Appendix A.5. As I went through each row, I realized that the first column – the existing solution – was not always worse than the second column – the new solution.

Then, I thought that I did not know of any method or decision making method or tool that could help in deciding which of the two was best. Therefore, I decided to initiate a quest that ended up in a scientific publication [86]. Before submission to a peer-reviewed journal, I did what I enjoy probably the most: ask the younger generations for their critical opinion. I had the chance to present it to my students in at least two different courses, and their impressions helped me find some weak angles; for more on this, see Section 2.7. Additionally, I used this method to persuade potential customers to purchase the product our small company was selling (see Table 3.1).

Interestingly, I did all of the above following the same time evolution of ingredients we discuss in this book (see for example Section 3.4.1). The only caveat is that I did all that almost intuitively. It took me several months to acquire the *knowledge* I was lacking. Then, I *persuaded* three other colleagues to accompany me in writing a scientific article, which luckily was published, without having to request funding or mass producing a product. With this book I hope I can make the bumpy ride more feasible for you: I am *empathizing with you*!

I have included some examples that will help you understand the *IF* method better and ways to teach it in Appendix A.5. More importantly, you may have already mastered the idea behind comparing between two options. But what if there are more than two alternatives? To keep the storyline of this book as short as possible, I invite you to check Appendix A.7 [95].

To make a long story short, the feeling that there was no tool or method to assist in taking decisions is what gave me the drive to read hundreds of publications. As a result, I developed the *IF* model I invite you to start using right away, not necessarily as an academic exercise, but also to tackle real-life problems!

Some more elaborated decision making methods do exist and are widely used in professional settings. To name only a few, Pareto analysis guides decisions concerning the most impactful steps, i. e., how to be economical with time and other resources; life cycle analysis (LCA) methods can guide decision making based upon sustainability perspectives; multiattribute utility theory [176] represents a broad area of approaches for decision making which accounts for the preferences of the decision maker.

Besides LCA methods, there are other metrics addressing, e. g., concerns in "green chemistry," such as atom economy (AE), the E-factor, reaction mass efficiency (RME) and process mass intensity (PMI or simply MI). These methods are sustained by a chemical algebra that picks one factor or combinations of several factors leading to quantitative analyses of chemical efficiency [47, 73, 122].

The *IF* method can be applied in cases where the simplifications outlined in Section 2.7 do not hold. For example, it has been adapted in many scenarios by other researchers who provide cases of how to deal with nonlinear changes in F [178, 160, 146, 71, 177].

The fact that other researchers have found value in the use of this method is a source of great joy, and I truly hope you can make good use of it one day.

2.8 Relation between A and *IF*

The applicability (A, equation (2.1)) and the intensification factor (IF, equation (2.2)) can be seen as proportional to each other, see equation (2.6), having a reinforcement or synergy effect. This means that higher *IF* values could be used as a predictor of higher chances of successful application of the solution at hand in reality. More knowledge accumulated with time should correspond to reliable arguments (i. e., factors F_n) that lead to more effective persuasiveness. Likewise, identifying the right set of factors can help empathize with other stakeholders, contributing to better outcomes. Knowing what "other factors" are limiting the applicability (that can be factors to calculate in the *IF* method) will help making contingency plans and increase the chances of success as well.

$$\begin{Bmatrix} F_1 \\ F_2 \\ F_3 \\ \vdots \\ F_n \end{Bmatrix} IF \propto A \begin{Bmatrix} K \propto t \\ P \\ E \\ OF \end{Bmatrix} \tag{2.6}$$

2.9 Changing K:P:E importance

The KPE framework does not indicate the depth of knowledge necessary here, as that will be dependent upon the complexity of the problem you or your team is being challenged with. When working as part of a group, persuasiveness may also have a role at this early stage when the problem is being defined.

When you are in the early phases of a project, such as in STEM-related assignments, it is common to have a problem statement. Having a well-defined problem helps to advance efficiently, and it may be redefined or updated as you make progress. Please note that in real life the starting point most probably comes from an empathic drive.[15] Actually, untrained people also identify problems and opportunities, and very often they are better at spotting opportunities outside the tunnel vision that experts have. This implies that E can be more important than K in certain moments.

Let us start assuming that for the STEM project we are in a loop at iteration $i = 1$, and you will probably iterate a few rounds, n. We refer to this stage as the *problem/opportunity identification* point shown in Figure 2.12.

As part of the problem statement, an idea might emerge in one person or a group of people, e. g., co-founders of a company. This idea is not necessarily a full solution – represented by the *light bulb* in Figure 2.12. At this stage, the importance of knowledge is highest, hence its larger proportion (or height) in the figure.

Arguably, only by knowing even vaguely about a problem that needs to be solved – *opportunity identification* – is it possible to generate applicable ideas worth following. This starting knowledge may be categorical or cross-functional, drawing from multiple disciplines. The opportunity needs to be clarified for every stakeholder or team member joining.

Before we move to the second stage in the KPE framework, we are faced with two intertwining paths:
1. to acquire missing knowledge; and/or
2. to persuade others to join or collaborate and address the complexity of the problem.

In the first option, acquiring knowledge will be easier if empathy can be used to help categorize what information is necessary or most impactful.

In the second option, empathy can be used to persuade others to join.

---

15 One of my PhD students does not agree with my statement because: "If empathy is defined as understanding the needs of society, then problem identification needs empathy more than knowledge. My father (untrained in the art of engineering or physical sciences) identified the need for an affordable de-humidifier device that could improve the ambient indoor conditions in my coastal city while also providing clean water. This did not require knowledge beyond the fact that the air was humid and that it was causing the discomfort." – I would argue that he knew about the problem from first hand, and therefore the need for empathy was not so strong. You can also identify a problem by having knowledge and decide not to do anything due to a lack of empathy.

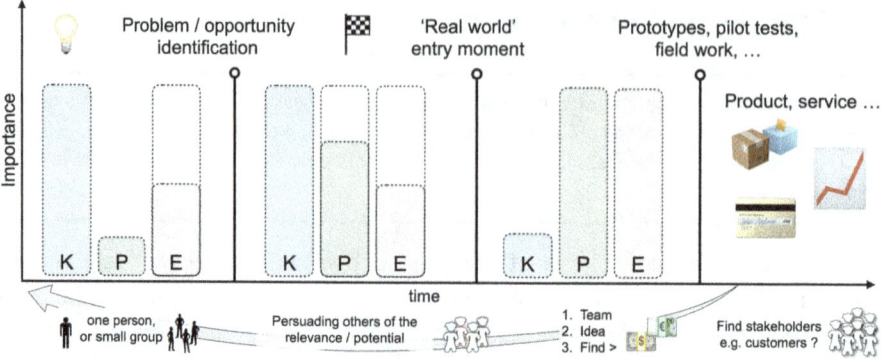

Figure 2.12: Schematic representation of the time evolution and the relative importance of knowledge, persuasiveness and empathy, at each point. The empty space defined by discontinuous lines indicates the possibility or convenience to have a higher value for each ingredient. Please note that I am not discussing ingredients or elements that are necessary such as creativity and critical thinking. The arrow pointing backwards indicates an iterative processes, which will be explored in Sections 2.9, 2.6 and 3.1 and Figures 2.12 and 3.2. Adapted from [96], Creative Commons CC-BY license.

Therefore, the importance of empathy increases at this stage compared to the previous stage. Empathy helps the lead engineer or innovating team to define if and how to invest resources, including risking time to advance further. Practically, having a certain degree of empathy can also help relate or connect better to those suffering the most from the problem – see the note on Dalton on page 36, Section 2.10. Ultimately, an adequate empathy level can act as a driver to continue forward in developing a solution.

After the problem and an opportunity to solve it are identified, the need for persuasiveness and empathy skills grows. This is important to scale up or speed up the process to reach the real-world entrance. This, sadly, is a point where most ideas or start-up companies tend to lose steam or fail completely. Logically, knowledge remains highly important, as the founder and the team must be prepared for many more scenarios that open up. The depth of the knowledge is more than what was needed while still defining the problem: this new relative importance ratio is seen in Figure 2.12. We could argue that K then restarts when a new phase is reached, and internal iterations between the first two steps may occur, while iterations or variations of the longer cycle are repeated.

Beyond the *real-world entry moment* shown in Figure 2.12, the ratio of importance shifts again, with persuasiveness and empathy dominating. You can think of this moment as when a company is created, or the first real actions are taken to start solving the problem, e. g., taking a loan or asking for support from "friends, fools and family."

If an idea or proposed solution does not resonate with more people besides the founder or core team, it is unlikely that the enterprise will succeed, or it may take longer to reach some tangible valorization or output.[16] It is here that a team's ability to be adaptable, agile and responsive to market conditions is especially crucial. There are many aspects which allow a team to encompass these traits, but a full discussion is beyond the scope of this book. However, an empathetic mindset can ensure that the team or inventor is not creating a product or service that nobody needs. For more on teams and leaders, see Sections 5.7.1 and 5.8.

In an ideal situation, a team will have been built around a polished idea at an early stage, i.e., it has been validated by several stakeholders or the market, for example, some early adopters. When I say "validated" I mean that you or your team members have reached outside your usual "thinking space." This can mean the office or lab where you regularly work, but also the division or departments in existing organizations.

However, the team will likely lack necessary resources to scale up. Acquisition of resources provides another reason that empathy and persuasiveness are key at this stage. Resources can take many forms besides money, such as physical space to perform work or institutional support for legal and logistical aspects. In any case, empathy is needed to make sure there is a clear connection between the solution, the stakeholders and the ultimate beneficiaries.

A simple example can be a small company needing money to build a *prototype*, so that the first customers can be approached to validate if there is some probability of someone willing to use it and establish its affordability.

It is typically just preceding or around this stage that start-ups enter what is called the "valley of death" [156]. Let us assume the team was able to overcome and cross this valley – typically death comes when there is not enough money to jump from idea or prototype to the next stages. What is then left is to scale up or expand the operations, i.e., *product or service*, of our hypothetical team and their brilliant solution to a societally relevant problem, and we can move to another iteration, $i + 1$. For more clarifications and examples on the changing importance of KPE, please see Section 3.4.

2.10 KPE's model case

This section is about an abstract or neutral case; if you are interested in real examples, please check Sections 3.4 and 6.2. Depending on the type of problem you want to solve and the specific professional or occupational field you are in, as well as your personal relation to the problem, you may define different ways to solve the problem. For example, it is generally easier to have knowledge about a problem when it affects you directly.

---

16 I provide a cautionary tale about founders and resonating ideas in Section 5.8.

 John Dalton (1766–1844) was an English chemist, physicist and meteorologist. He is known for his work on the atomic theory and human optics.

> "One reason why Dalton became well known, (...), was that he was colour blind. This condition had not previously been recognized, but Dalton came to realize that he could not see colours the same way most other people could, and found that his brother was affected in the same way. Blue and pink, in particular, were indistinguishable to both of them. On 31 October 1794 Dalton read a paper to the Manchester Literary and Philosophical Society describing his detailed analysis condition, which soon became known as Daltonism (a name still used in some parts of the world)."
>
> The Scientists: A History of Science Told Through the Lives of Its Greatest Inventors
> by John Gribbin [104]

If interaction with the problem is indirect, different tools are needed to acquire valuable knowledge, e. g., marketing actions and customer surveys. Let us think of this hypothetical scenario, stripped of any specific information so that the bare logic behind the proposed KPE framework becomes clearer.

- We start again in the proposed order (KPE), at $i = 1$, where knowledge pertaining to a problem is acquired (see Figure 2.12). At this stage, an isolated solution will face more inertia and rejection from stakeholders; therefore, the lead engineer will need to clarify what are the problem's drivers, what constraints exist, etc.
- To move forward, the engineer will likely need to develop an ability to convince persuasively – there are difficulties associated with not having a convincing track record or professional experiences, and this specific problem is shortly discussed in Section 5.11.1. Perhaps more importantly, the engineer will need tools to demonstrate how the new idea compares to current alternatives; here, the *IF* method could be utilized, as given in Section 2.7, and other examples are given in Sections 3.4 and 6.2.
- Once the entrepreneurial engineer or innovation team decides to move forward, assuming they have the level of empathy needed to understand the problem and relevance of their solution, they must also empathize with other stakeholders. At this stage, the entrepreneur or team will need to convert a solution into a value proposition that can be of interest to the stakeholders they are communicating with. This empathy can be authentic (trained or innate), altruistic and/or transactional (see more in Section 4.4.4).
- It should be noted that the KPE framework involves multiway transactions. It is important to acquire knowledge, but relevant portions must also be transferred to the rest of the team and stakeholders. Team members will require varying levels of persuasiveness depending upon their role, but *may need to be persuaded as well*, should a fact or decision arise that does not align with their original views. As detailed in Section 2.9, the importance of each ingredient will grow or shrink depending on which stage of the process the team or solution is at: between $i \geq 1$ and $i \leq \infty$. More on the importance of knowing when to follow or lead can be found in Section 5.7.

3 Give me facts...

Reading time ~ 40 min

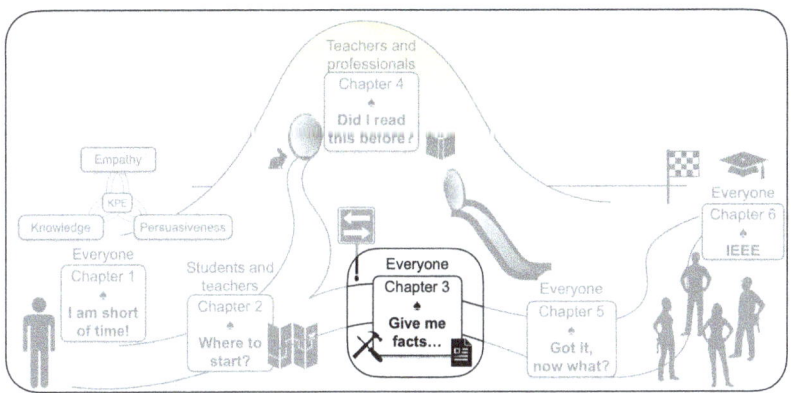

Bit: Where you get the tools to work

> "Nullius in verba" ~ "Take nobody's word for it"
> Roughly saying that you should verify statements by relying on evidence and facts.
>
> The Royal Society, one of the first scientific societies

What use would this book have if it does not provide you with some steps to try out? Or how could you get inspiration if I did not share some examples that have helped other people in fulfilling their dream to innovate or solve challenging problems? The first answer is presented next; for the second, carry on to Section 3.4.

3.1 Innovation guide

Now that we have covered in the previous chapters a foundation built on knowledge, persuasiveness and empathy, the next generation of innovating engineers should be well equipped to tackle innovation processes. Here I introduce six steps to innovate which bring together the principles of the prior chapters (see Figure 3.1). These steps are interrelated with KPE and the relationship will be explained in the subsequent sections, backed up by several case scenarios.

1. Defining magnitude of problem
2. Devise a first solution, milestones, reflect on limitations
3. Define metrics
4. Gather information, assemble team
5. Break it down
6. Calculating and deciding

Figure 3.1: Step-by-step procedure that can assist in the decision making of when and how to innovate or change a given process or product. This sequence can be seen as minicycles introduced in Section 4.4, Figure 4.2. Adapted from [96], Creative Commons CC-BY license.

I. Defining the magnitude of the problem

If we assume a typical STEM-related problem from academic or industry settings, in early stages, *knowledge* is likely the most important of the three ingredients highlighted in this book (see Figure 2.12). As discussed in Section 2.9, in real-life scenarios, an engineer with a good empathy level or an untrained empathic person can be suited to initiate an innovation journey.

Clearly defining the total number of people impacted by the problem and eventual solution, identifying the most impacted subgroups, obtaining a monetary estimate of the impact, identifying additional (nonmonetary) problems that the main problem can lead to and understanding the context are important aspects.

Here, using a framework based on political, economic, social, technological, environmental and legal (PESTEL) factors can be useful [137]. PESTEL is used to analyze and monitor the macroenvironmental (external marketing environment) factors. This is typically seen in the context of the impact it may have on an organization, company or industry. These factors can also be used to identify threats and weaknesses, a popular analysis known as SWOT [22]: strengths, weaknesses, opportunities and threats.

However, additional factors such as demographic, intercultural, ethical and ecological factors can also be added. *Empathy* might have some use at this initial stage in fully defining the extent of the impact a problem has on some people, but it may also cause one to focus too much on a subgroup of people and narrow too much the scope (see Figure 3.2). *Persuasiveness* could be seen in a "self-persuading" dialogue that can lead to convincing you to continue to the next step or stop.

II. Devise a first solution, define milestones, reflect on limitations

Subdivide the process of creating a first-attempt solution into achievable milestones. Plan steps that will allow incremental achievements towards creating a comprehensive solution; for example a timeline and mind map as presented in Figure 5.2. Simultaneously, you have to budget time and resources for reflection, analysis and feedback to incorporate changes and adapt as new information is obtained. See Chapter 5 for some advice. This initial solution can be what is commonly known as minimal viable product or prototype (MVP).

Clearly define the limitations of each step to attempt to predict what you might not know at a certain stage; doing so comprehensively will not be possible without extensive prior experience. But adopting a strategy that allows for re-evaluation lowers the risk of missing information that can significantly change the direction one is going. Clearly define a "go/no-go step" or time period for each projected stage in your innovation effort.

During this stage, you should evaluate the time and resources you have at your disposal to continue going and weigh these against the probability of success or risk of failure (see Section 5.11). If possible, establish contingency plans if there are several "go" or "no-go" options that can salvage part of the achievements made up to that point. *A "no-go" does not have to mean a hard stop of the work done so far, but may indicate that a pivot towards a new end-goal is necessary.*

At this stage, knowledge and continuous information gathering is still likely the most important out of KPE (see Figure 3.2). Empathy will begin to play a larger role when defining "go" vs "no-go" scenarios, particularly when working in a team. Be aware that you are now risking the time and resources of other stakeholders in your entrepreneurial pursuit. This all happens while trying to match the market demands and managing expectations from the group for whom you are developing a solution. Thus, sizeable levels of persuasiveness may be needed or desirable.

III. Define metrics

Here comes when we need to select what quantitative or qualitative aspects are important to reach each milestone. The selection allows for focus and makes it clear what information and knowledge is needed most urgently. Most importantly, define several metrics or factors that can be used to evaluate whether the problem was mitigated by a solution; these factors can be defined even without a specific solution already developed.

Such factors can be used to define a *before* and an *after* the introduction of each solution. The process to approximate a quantification for this evaluation is explained in Section 2.7 and expanded in [86], termed the *IF* method. It can also increase applicability by effective mitigation of other factors (see equation (2.1)). If there is already a team assembled, its internal communication can be efficient when persuasiveness is well done among team members. That clear communication helps with informing or persuading stakeholders concerned with the problem.

If you are able to build a good metric's system that can be understood by all stakeholders, it will be easier to achieve more empathy among the stakeholders and within the team.

Figure 3.2 shows some suggested levels of the KPE at each step in an innovation journey. These levels are indicative and should be taken as relative to each other.

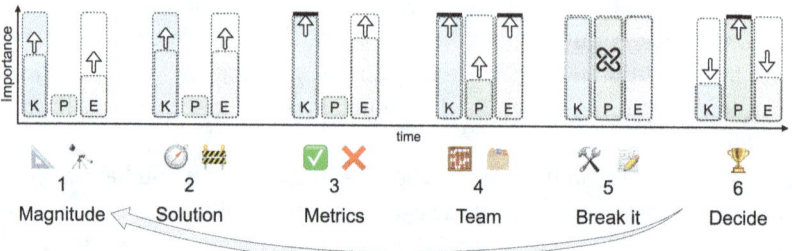

Figure 3.2: How KPE ratios change with time as you execute the innovation steps. Arrows pointing up indicate that an increase in each ingredient is desirable. Arrows at the top pointing at the thicker black line mean that topped up ingredients are ideal, whereas those pointing down indicate less relevance. The entangled symbol "crazy" in Step 5 indicates mixed and maxed up ingredients during a creative process. The arrow pointing to Step 1 implies the cyclic or iterative nature of innovating processes. Adapted from [96], Creative Commons CC-BY license.

IV. Gather information, assemble a team

Collect the additional data that were not available or in your focus from Step 1. Assign values to the *before* and *after* corresponding to the *IF* method introduced in the previous step. Data collection can occur in numerous ways: from seeking out published data to carrying out experiments or preferably building a prototype solution if possible.

If you are lacking some knowledge, e. g., exact values for each factor cannot be obtained, estimations can be used, but the corresponding assumptions should be clearly defined so that the limitations of these estimates are understood and kept in mind. This transparency is needed for effective and honest persuasiveness.

At this stage, with metrics defined, the limitations to gather information of the current person(s) working on the solution may become apparent. It may be beneficial to then bring others onto the team. Here, the relative importance of persuasiveness (and thus relationship building) will increase significantly compared to prior steps. Empathy levels are also expected to be topped up, because the metrics and magnitude of the problem are clearer than during the earlier steps.

V. Break it down

This step borrows its name from music – notably, jazz – where each musician is given room to improvise and do what they are best at. This is perhaps the most inaccessible of all innovation processes, as it relies upon the creativity of individuals or of a collective group. It is where each team member contributes with their own strength. A good creative team benefits from having leaders and members able to delegate specific activities and knowing when each member can perform in the best possible way.

The symbol in Figure 3.2 represents the "crazy" entanglement of the three KPE ingredients during this step. This step is strongly dependent on the specific subfield of engineering and the actual problem and context. In Section 3.2, I provide one particular example on how this can be done with the SEST template. The "theory of inventive problem solving" (TRIZ) could fit here as well (https://en.wikipedia.org/wiki/TRIZ), and you can find more elements about it in Section 4.8.

VI. Calculating and deciding

This can be seen as a recurring step which should come whenever significant progress toward a milestone has been made or highly impactful new information has been obtained. It is a moment where a decision to proceed or not is made, referenced in Step 2 as the "go or no-go" point. Based on the discussion from Section 2.7, you may use the calculated IF_{total} value to assess if is it larger than 1; if so, it may make sense to continue.

Moreover, you can use the applicability concept (see equation (2.6)) together with the IF to assess with the team and stakeholders what is the most desirable step forward. Here, persuasiveness is topped up because there may be people not fully convinced to move forward, even when knowledge and empathy may be decreasing.

We could add a substep for testing the assumption or hypothesis, for example to validate it by asking experts or users of a given market, testing a prototype device, e. g., a lean start-up loop.

I know that these steps sound very vague at this moment; therefore, before we move to more specific examples where these steps will be properly explained, let me bring back the simpler case of the city trip from page 30 and the three friends from Figure 2.11. This is very similar to how I would explain it in the classroom:

| You and two friends are going on vacation, but cannot agree on the location to visit and the duration of the visit, and you have to decide before making the necessary arrangements. You all have managed to agree on a total budget and have asked each other to prepare a list of important aspects and things you are absolutely not planning on doing. For example, you know what means of transportation may be available and possess partial knowledge about some of the cities you could visit.

II   The solution will be approached by defining a bunch of partial solutions, for example, ruling out traveling by boat if one of you does not feel comfortable, or if the budget does not allow a specific destination, you can start scrapping from the list of potential solutions. Here, you all will be asking around, checking travel brochures, with the right friendly empathic and persuasive means towards effectively defining your "boundary conditions."

III   Then, you will compile a short list of the factors that are important for each of you, agree which of them need to be put together and compare the value of each factor that relates to each solution or alternative. *This is related to K.* Examples include ticket prices of different transportation means, the cost of a single or shared room, visa requirements, etc. You may not know the exact or even any approximate price or duration of visa request, but that comes in the next step.

IV   You start to populate the values of each factor for each of your alternatives. Let us assume each of you wants to go to a different city, same continent, but two of you want a different country, one of them requiring visas. You already have the team – but maybe you find out one of your friends recently visited one of the locations you are targeting, and you invite this friend to assist you, but he is not joining for the trip.

V   This example we are discussing is very simple from the "break it down" aspect, because already each factor gives a qualitative or quantitative indication, and eventually you will have to decide on what suits this small team. As you will see in the examples later in this book, you may need specific tools or models to help you (Section 3.4 and onwards). You could also visit a travel agency to use their "secret sauce"; often they can make (innovative) arrangements that you may have overlooked.

VI   Then, you put all your numbers and considerations and calculate the total expenses, risks of failing to get the visa on time before purchasing cheap tickets, etc., and a decision should be within reach ... I hope for you!

You can see a simple calculation based on this example in Table 2.2, page 30.

I need to stress again, as in several parts of this book, most of the graphics shown here have some sort of "linear" timeline. This example is purely for educational purposes because learning and innovating are far from a linear process; see Figure 5.1 for characteristic iterations or loops to clarify a problem.

3.2 Inspiration template to innovate

In the context of education challenges, a group of collaborators and I proposed that innovation can be a means of achieving sustainability – primarily in the chemical industry, but applicable elsewhere. It can also potentially contribute to the United Nations Sustainable Development Goals (UNSDGs) [93, 94]. The best way we have identified to innovate – or intensify chemical processes – was introduced earlier as a classification by four different approaches that either individually or combined can achieve various degrees of innovation or intensification [172]. It is related to modern chemical engineering because of its goal of reaching more sustainable and efficient ways to manufacture chemical products.

> *SEST template*: Different domains to innovate, borrowed from process intensification, used in the upcoming sections and depicted in Figure 3.3:
> - **S**tructure – spatial domain,
> - **E**nergy – thermodynamic domain,
> - **S**ynergy – functional domain, and
> - **T**ime – temporal domain.
>
> These four approaches can be used in mechanical and environmental engineering, as well as other fields of knowledge such as business operations (see Section 3.4 for examples). Basically, what this division of approaches provides is a set of concepts that guide "what to change." In fact, applying these four approaches is an example of what I suggest in "break it down," Step 5, in the innovation guide presented at the beginning of Section 3.1. Consider it a secret sauce cooked by chemical engineers, but please, feel free to use it elsewhere, like me, to spice up your work!

If you are not a chemical engineering student, do not panic when reading this part. With a bit of abstract thinking – and examples I will provide, e. g., not related to chemistry in Sections 3.4.2 and 6.2 – you will be able to find analogies with other STEM disciplines.

Structure You can innovate almost anything you can think of by *changing the geometry* or interconnection of parts. You can alter the shape of a mechanical component by making it lighter, e. g., taking away excedent parts, like drilling holes in a solid beam or frame or replacing the materials. But it can also be achieved by changing the order of a logistics chain or the order in programming sequences in a software program.

Energy Examples of using the energy approach include changing the power source needed in a mechatronic device from wall-charging to a portable battery or changing the energy source from conventional fuels to solar. But you could also make it more abstract; *think of how the energy is used by the device, in a direct way or via transformations*. For example, there are spark-ignited devices used to inject without needles during medical treatments that can be replaced by a spring-loaded system [157].

Synergy I feel that perhaps the simplest way to explain the concept of synergy, plainly speaking, is making a cocktail: *mixing elements* that provide a better result than the sum of separate parts. Other examples include conventional architecture or the design of any system. Here, I always think of going loose and see what happens. In practice, this can be seen when a heat exchanger and a reactor are combined into a new unit.

Time This can be interpreted as altering the *temporal component*, e. g., changing the flow rate or operation temperature in a reactor. In sonochemistry, for example, this can be done by adjusting the frequency of the ultrasound: lower frequencies have different effects on a given experiment than higher ones (see Section 3.4.1). For those with a chemical background, you can imagine the benefits of having a reactor that has the ability to reverse flow conditions and restore the properties

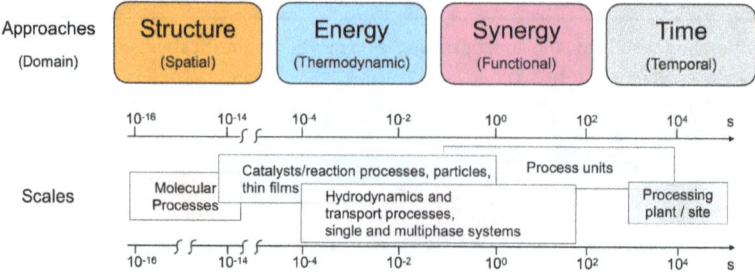

Figure 3.3: A schematic representation of the approaches and related scales belonging to the process intensification framework. The approaches to successfully intensify (or innovate) a process can be grouped in four domains: spatial, thermodynamic, functional and temporal, SEST. These approaches can be applied on the relevant time and length scales shown on the lower bar, ranging from the molecular level to the size of a chemical plant. Reprinted and adapted with permission from Structure, Energy, Synergy, Time – The Fundamentals of Process Intensification. Copyright 2009, American Chemical Society [172].

of a catalytic filter. If you are not into chemistry, but have seen cars fitted with a cleaner exhaust that filters toxic components, that is probably a catalytic filter.

In heterogeneous catalysis, there are the "Langmuir" and "Mittasch" approaches [83]:
- Langmuir approach: using single crystals to test simple low-pressure behavior of gases on a surface. This is the bottom-up approach to catalyst design.
- Mittasch approach: testing at high temperature/pressure with thousands of catalysts and finding the best formulation through testing.

Both approaches are valid, as the bottom-up approach helps us to understand how to construct the best possible catalyst with the current "knowledge." However, the Mittasch approach also points to results that may not have been deduced from a bottom-up approach. For example, a multi-promoted iron-based catalyst for ammonia synthesis. It is as if the Langmuir approach looks more into the science and what exactly happens here (expanding your knowledge), whereas the Mittasch approach is more based on application and trial-and-error to see which catalyst is the best, and when you do not know sufficiently enough about the case in study.

You can see in Figure 3.3 the SEST template and the corresponding length scales where you can innovate. It goes accompanied by the main goals of process intensification (not shown in the figure):
1. maximize the effectiveness of intra- and intermolecular events;
2. give each molecule the same processing experience;
3. optimize the driving forces at every scale and maximize the specific surface area; and
4. maximize the synergistic effects from partial processes.

Explaining the goals and the interrelations between the approaches is, unfortunately, beyond the scope of this book, but please dig into the literature provided if you want to know more.

Nevertheless, I will use a particular example related to microfluidics that we will explore in the following chapters. Microfluidics can address all these goals, particularly because chemical engineers have learned in the past decades that "larger" equipment is not always necessary to reach an economy of scale [97].

> "(...) from a scientific perspective the truly revolutionary character of the Industrial Revolution was the dramatic change from an open system where energy is supplied externally by the sun to a closed system where energy is supplied internally by fossil fuel. This is a fundamental systemic change with huge thermodynamic consequences, because in a closed system the Second Law of Thermodynamics and its requirement that entropy always increases strictly applies.
> We 'progressed' from an external, reliable, and constant source of energy to one that is internal, unreliable, and variable. (...) An example of the consequences of the Second Law is the warming of the atmosphere due to the release of energy stored underground in fossil fuels onto the surface of the planet. This is strongly enhanced by the production of gases such as carbon dioxide and methane as entropic by-products from burning these fuels, leading to the well-known greenhouse effect in which heat gets trapped in the atmosphere."
>
> Geoffrey West in Scale [180]

3.3 Size matters

To give a flavor of innovation in research settings, I will share an interesting message my collaborators and I published a few years ago. We realized that the experts working in "Energy" had to be *persuaded* about the advantages of microfluidic reactors, such as their small size.

In this book you will find several parts where I talk about microfluidics, but in fact, we should focus on a slightly wider definition of microsystems as "devices with at least one characteristic dimension in the micrometer scale" [136]. These devices could operate either in a continuous fluid flow regime (microfluidics), such as in the case of electrolyzers, fuel cells and flow batteries, or in a static regime, such as batteries and supercapacitors.

We provoked the readership of the journal with the following questions:
- Is there really "space" for "small" technologies in the energy challenges faced by our society?
- What do microsystems have to offer that large-scale technologies have not already provided?

When you miniaturize a reactor or fluidic system by decreasing a characteristic length L (e. g., channel width, distance between electrodes, etc.), you increase the gradients, such as concentration or temperature, which is typically desired. From a safety perspective, handling small volumes can be desirable to avoid accidents or spillage and reduce waste generation due to wrong outcome of a given reaction or process. Moreover, fluid flow in microchannels tends to be in the laminar regime, which is easier to control and predict.

Lastly, microdevices have a large surface-to-volume ratio (m^2/m^3), another desirable feature when trying to control processes dominated by surface phenomena, such as heat and mass transfer. To convince you about these advantages when going "small" or working at the mesoscale, please see Figure 3.4, top.

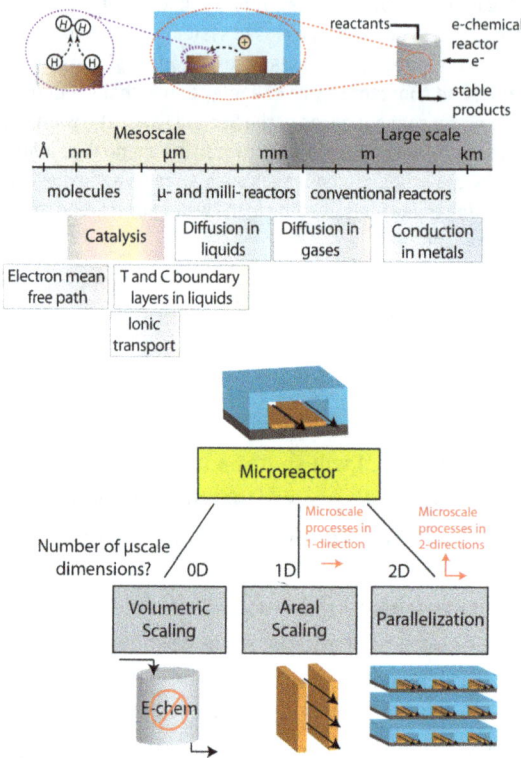

Figure 3.4: Top: Diagram illustrating the main processes present in electrochemical energy conversion devices. From left to right, processes are presented in order of increased limiting scales. Please note that most processes lie within the mesoscale. Bottom: Proposed strategies for increasing the throughput of microfluidic reactors, also known as scaling. The conventional method for homogeneous chemical processes where the dimensionality of the reactor does not affect the reactions (0D) is volumetric scaling. Areal scaling is preferred if the limiting transport processes relying on microscale path lengths can be carried out in one dimension (1D). Microdevices can be parallelized when the system involves more than one dimension in the microscale (≥2D). These images were taken from an Open Access Article licensed under a Creative Commons Attribution-Non Commercial 3.0 Unported Licence [136].

Logically, the first question that comes to mind is, how can all these benefits observed at such small scales be extrapolated for the production of large quantities of chemicals or energy conversion? In yet another example of how engineers manage to borrow from different fields, microfluidics researchers have benefited enormously from the fast development in the semiconductor industry by using the same technology to fabricate electronic microcomponents, but for different aims.

In the last three decades, microfluidics grew from an odd scientific field into successful companies that operate worldwide. Nowadays, microsystems are made of several types of materials and used in different applications, from pharmaceutics to food and energy [97, 48]. One of the classical answers I give when people ask how it is possible to produce more with microsystems is by *numbering up*, or parallelizing an existing microsystem (see Figure 3.4, bottom).

To fabricate these microsystems, we have inexpensive manufacturing processes available that are cost-effective for large-scale microfluidic reactors and new fabrication techniques such as high-resolution additive manufacturing to innovate even further. Unfortunately, we cannot address all these interesting aspects, but I invite you to read further in the cited literature above.

Despite the advantages of microchemical systems and successful applications mainly in chemical engineering research, mainly pioneering companies have been able to capitalize on this innovation. A very well-known reason for its limited adoption in commercial environments and at larger scales is the fact that small channels clog. Then, the *knowledge* to avoid this has become instrumental in facilitating the deployment of these microfluidic systems to solve societal-relevant problems, for example, which materials and geometries minimize particles' attachment to surfaces that lead to clogging the microchannels.

Figure 3.5 shows a comprehensive diagram with terms and concepts we have introduced until now. I hope that you will understand how synergy, understood in the framework of process intensification, has the largest relevance.

3.4 First-hand case analyses

In what follows, I begin with two personal stories of an innovation trajectory that I undertook before understanding all that I am writing in this book. Please, also check the "journeys" of other empathic entrepreneurial engineers in Section 6.2. You will find complementary examples with references to the innovation process guide we discussed in Section 3.1 and KPE.

3.4.1 From PhD student to entrepreneur

The basic structure of this section in the book emerged after being invited to contribute to the journal *The Cuban Scientist* with a two-page journey from my PhD project that

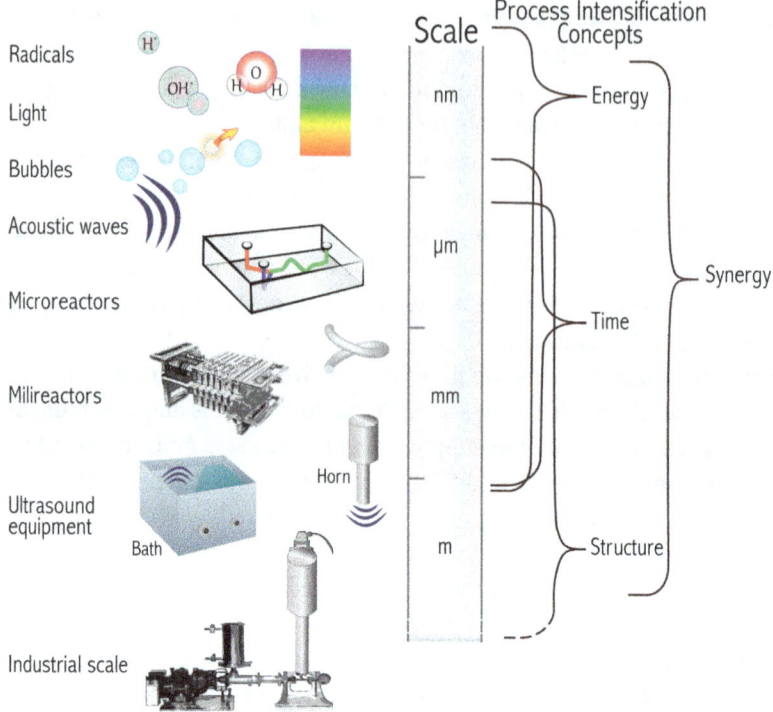

Figure 3.5: Here you can see a challenging picture illustrating why the lack of knowledge required at each "scale" from an expert focusing on another specific scale can limit the utilization of sonochemistry at a worldwide level. Observe the relative sizes of the items described on the left, increasing from top to bottom. Notice also how in the right part, different sizes can be related to the process intensification SEST concepts having the greatest influence. For example, we cannot easily change structure below a given nanometer scale, but we can use synergy at all scales. Taken with permission from [97].

turned into a spin-off company on ultrasonic cleaning and advanced chemical processes. The basics of sonochemistry and process intensification are also introduced here [89]. Please, keep in mind the *steps* shown in boldface in Section 3.1 in the narrative below.

I started my PhD project in 2007 with the aim to increase the efficiency of a specific type of chemical reactors using cavitation for water remediation – *Step 1 – defining problem magnitude*. This initial step was already defined by the supervisors of the project, but in my experience, it worked best when I, as PhD student, updated the knowledge[1] and refined the focus.

---

[1] It is common that a PhD student starts a project months after a research proposal was written.

At that point, I found in the literature many energy sources for advanced oxidation (for water remediation) to choose from, e. g., hydrodynamic cavitation, ultrasonic cavitation (sonochemistry), catalysts activated with ultraviolet light, the addition of specific chemicals and the combination of some of them – *Step 2 – devise solution, milestones, limitations*. All of the above would be considered part of the *thermodynamic domain* approach in the SEST framework introduced in Section 3.2. During the expansion phase – lower pressure – of a sound wave, bubbles can be formed (see Figure 3.6).

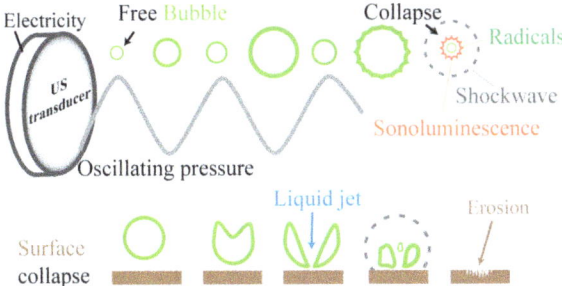

Figure 3.6: A free bubble and a bubble close to a wall behave differently as they oscillate driven by pressure variations produced by an ultrasound transducer. At the moment of collapse, radicals, shockwaves and light emission (sonoluminescence) can be measured. When near a surface or another bubble, jetting can also occur. Repeated collapse events against a surface can lead to erosion of surfaces. Taken from [97].

In the case of ultrasound, different studies tried to find an optimal frequency to achieve better results, which would be part of the *temporal domain*. Some other attempts included the combination of different energy sources and frequencies, which corresponds to a *functional domain* approach. The metrics for success I defined – *Step 3* – were based on counting the number of bubbles and correlating it with the amount of chemical radicals produced in total.

The most known hurdle for ultrasound and sonochemistry to be widely employed as a useful tool has been the very low-energy efficiency values. Acoustic transducers transform electrical power into mechanical energy which in turn is transmitted to the liquid. As a result, the energy goes partially to cavitation, or bubble formation, and the rest heats the whole system.

Consequently, not all of the energy produces the desired chemical and physical effects, and it is difficult to establish a robust energy balance because the quantities involved are not always easy to measure. I then defined the sonochemical efficiency (or yield) as X_{US} = measured effect divided by input power:

$$X_{US} = \frac{\Delta H (\Delta N_{rad}/\Delta t)}{P_{US}}, \qquad (3.1)$$

where ΔH is the energy required for the formation of OH· radicals, which is equal to the enthalpy of formation of the chemical reaction with a value of 5.1 eV per molecule,

$$H_2O \underset{}{\overset{\Delta H=5.1\,eV}{\rightleftharpoons}} OH^{\cdot} + H^{\cdot}, \tag{3.2}$$

and P_{US} is the electric power absorbed by the transducer, determined from the measured voltage and current and their phase difference. On average, the values reported for OH· radicals are in the order of $X_{US} \sim \mathcal{O}(10^{-6})$, which are very small for commercial interests.

Ultrasonic cavitation and its effects are notoriously difficult to reproduce because bubbles are in general created from impurities inside the reactor. The random defects on the walls (crevices) or dissolved solid particles are efficient traps for gas nuclei, as introduced in the Knowledge question on p. 10 (see Figure 3.7).

What made my research efforts really innovative and successful – *Step 4 – gather information and team* – was that I chose the *spatial domain* approach by controlling the place where bubbles are created and later implode due to pressure variations [90] (see Knowledge assignment on page 10).

I also had the opportunity to collaborate freely, and I kind of assembled a dream team of international collaborators for all my publications. This interdisciplinary collaboration between microfabrication experts, physicists and chemists led to the design of a new microfluidic sonochemical reactor, which we modeled and tested under laboratory conditions: my thesis cover has a very cool design, and you can find more about our results here [88]. The idea was so innovative that we persuaded the Intellectual transfer department to apply successfully for a patent [173], which became the BuBble Bag, as shown in Figure 3.8.

The "a-ha!" moment of my PhD came while attending a lecture given by Detlef Lohse in 2008. Detlef has been fascinated by bubbles for decades, and his work on fundamental science with relevant applications for society has stimulated generations of scientists and entrepreneurs. You can find his "personal scientific bubble journey" solving many puzzles elsewhere [127].

His 2008 lecture contained several beautiful experiments, but two in particular caught my attention. The first was performed in 1997 [126], when I was just 16 years old and had no idea about the formation and collapse of bubbles – cavitation – other than blowing soap bubbles as a child.

Detlef and collaborators trapped individual bubbles made of different gases at the center of a water-filled flask by using a sort of audio speakers glued to its sides. The speakers, made of piezoelectric materials, induced alternating pressure values that generated forces keeping the bubble in its place and made it grow and collapse repeatedly. When the pressure is low, the bubble expands, and when the pressure increases, the bubble shrinks.

What was fascinating to me is that under certain conditions, researchers had observed light coming out of the bubble, but the reasons were obscure. Detlef's work proposed a theory that clarified the dependence of inert gas content within the bubble, by combining principles from sonochemistry and hydrodynamic stability.

The second inspiring experiment was one where he showed the control over bubble formation induced by a shockwave traveling over a surface. The trick was to indent the surface with small artificial crevices or defects to stabilize bubbles [57]. You can compare it to the glass of water Knowledge question (see Section 2.3.3).

3.4 First-hand case analyses — 51

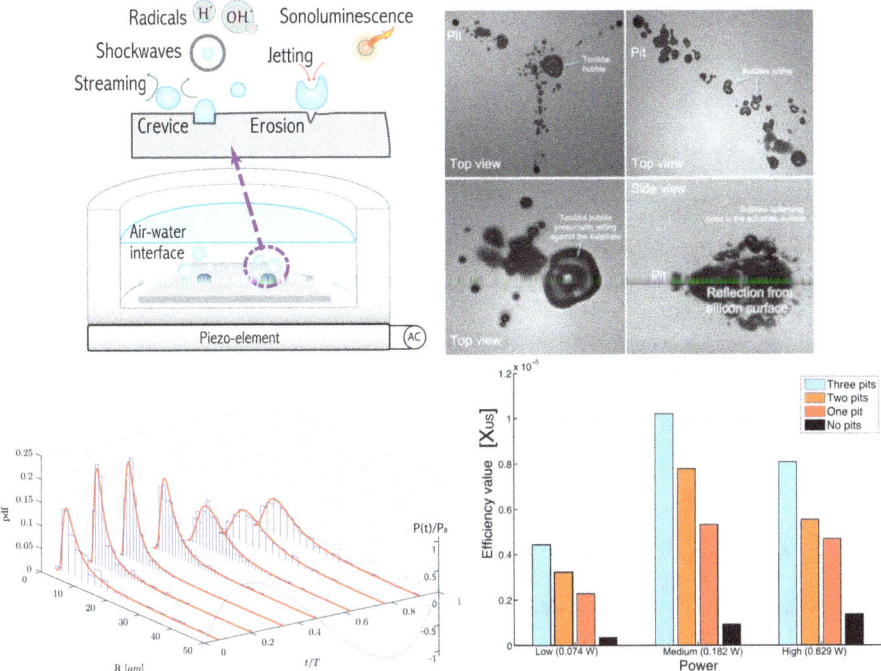

Figure 3.7: Top: (left) Diagram depicting the main phenomena associated to acoustic cavitation: radical production, shockwaves, sonoluminescence, jetting and erosion. Bubbles are formed from crevices existing in the walls of the reactor or dissolved solid particles. At the bottom, we see a tailor-made ultrasonic setup following the carbonated water example in the question on page 10. This reactor features a transducer (piezoelement) glued at the bottom of the container which converts electricity into mechanical oscillations that are transferred to the liquid contained in the reactor. This configuration has been used for several microfluidics studies. Taken from [97]. (right) Short-time exposure image at high power settings (0.981 W) showing deformed bubbles and jetting phenomena that can cause energy losses, affecting the overall efficiency. Bottom: (left) Bubble size distribution histograms at a power of 0.981 W for three pits. The axis to the extreme right represents the normalized pressure for the acoustic cycle. (right) Experimental efficiency values (X_{US}) for different numbers of crevices and different US powers calculated from equation (3.1) [92]. The presence of crevices increases the efficiency for each power. As the power is increased from low to medium the trends increase for any number of crevices, except for high power. This change in efficiency has to do with different ways in which the bubbles grow and collapse.

I was clearly impressed by the beauty of bubbles and I gained insight into untapped engineering solutions. Therefore, after his lecture finished, I approached Detlef and asked if he thought I could combine his two experiments with the aim of measuring chemical effects from those bubbles under continuous ultrasound driving.[2]

---

[2] Ultrasound refers to the use of sound frequencies above the audible limit of human hearing, typically larger than 20 kHz.

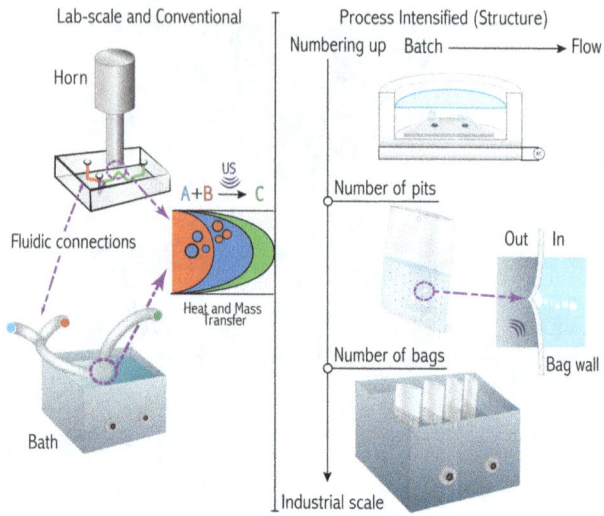

Figure 3.8: The left pane shows the conventional and lab-scale uses of traditional ultrasonic devices, horns and baths, used to promote reactions. To the right you can see what I think is a good example of scaling up a microsonochemical reactor by increasing its volume and numbering up the amount of crevices and the BuBble Bag. With permission from [97].

Months later, to my surprise, instead of having the trapped bubbles "do the chemistry" as I initially wanted it, beautiful "firework-like" bubbles were visible through my microscope eyepiece connected to a fast camera; see the 2012 Gallery of Fluid Motion – Taming acoustic cavitation (https://www.youtube.com/watch?v=kzJVw9HmKyg&t=92s).

That conversation has proven to be the most successful and long-lasting collaboration in my professional journey. I have learned many things from Detlef, for example, to always be curious, no matter if you work on fundamental and applied problems. He says that to find solutions, we must watch, listen, and be open, but make sure you work on problems that you enjoy trying to solve.

You can find more of his teaching and inspiration in his paper entitled 'Bubble puzzles: from fundamentals to applications' [127].

As several examples in this book show, this crevice trick became the foundation of my PhD and the spin-off company BuBclean [89]. This story shows the importance of following different fields of work, because you can gain inspiration to solve your own problems from many unexpected events.

It is important to imagine each bubble as a reactor in itself. Then, an undetermined number of bubbles of different sizes will be created whether you use an ultrasonic bath or a horn. Each bubble collapse will result in different temperatures and pressures inside of the bubble that depend on the different sizes and the collapsing velocity. This variability yields different products or effects, which means that the chemical or physical effects have a broad distribution of values, just because all bubble reactors behave differently, and ultimately leads to loss of controllability.

If we follow the "inspiration" SEST template of process intensification, Section 3.2, what I did was to change the *structure* of a batch ultrasonic reactor by micromachining artificial crevices of ca. 30 µm diameter and 10 µm depth onto the surface of silicon substrates. This modification trapped a determined number of bubbles (see Figures 3.7 and 3.8). During sonication, these bubbles pinned to the crevices serve as a seeding gas volume, forming lots of smaller bubbles containing a mixture of gas and water vapor; you can see more in the 2012 Gallery of Fluid Motion – Taming acoustic cavitation (https://www.youtube.com/watch?v=kzJVw9HmKyg&t=92s).

After careful experimental precautions, using fast cameras and with the help of close collaborators I measured the number of bubbles and the radicals created. Dividing the radical production per cycle by the average number of bubbles per cycle (see equation (3.1)) gave bubble radical production with efficiencies higher when the crevices were used (see Figure 3.7).

KPE relation with my PhD

Before nearing the end of my PhD, it never occurred to me to become an entrepreneur in the most popularly known sense – to start a company – nor did I know the extensive literature about it [158]. However, during experiments in the lab, not only did I gain *knowledge*, I also met a colleague who later became co-founder of a spin-off from the University of Twente, BuBclean (http://www.bubclean.nl) (see Figure 3.9). In all these actions, you can imagine how important it was to assemble the right team for each publication or step (see more in Section 5.10). We used *persuasiveness* to bring together different experts from almost orthogonal fields, which means "not having much to do with each other." Lastly, we grew *empathy* as we interacted with people who understood or suffered from the "problem" we set out to solve. A short recount follows next.

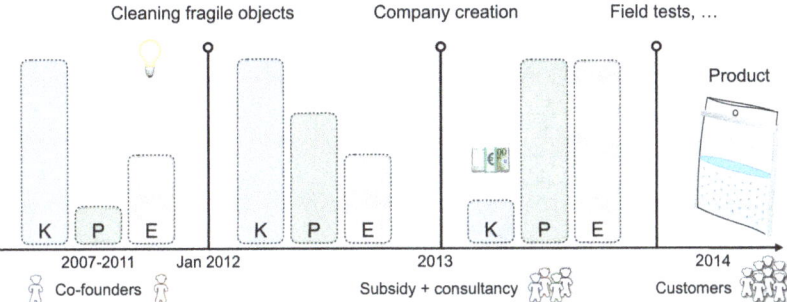

Figure 3.9: Evolution of an innovation trajectory and the role of KPE, from the scientific research phases up to ongoing commercialization. The "product" is a bag with a modified internal surface that entraps bubbles (spatial domain), and these bubbles generate more effects needed for cleaning and water purification, among other uses, for more information see http://www.bubclean.nl/bubble-bags-2/.

Before obtaining our PhD degree, my co-founder Bram Verhaagen and I had relatively vague ideas of what relevance to society our work had, and it was mainly in the framework of each other's PhD research topic. In my case I framed the use of these bubbles for water purification, and my co-founder was demonstrating how root canal treatments in dentistry could be improved with ultrasound. We independently honed our problem solving skills during approximately four years of research each, which, combined, allowed us to break down the challenges into manageable bits – *Step 5 – break it down*.[3]

Without entering into details, at some point we decided to explore the possibility of using our combined knowledge to solve a "bigger or real-life problem" around January 2012 (see Figures 4.3 and 7.1).[4]

From then on, we realized that we needed to *persuade* our supervisors and faculty to give us the opportunity to apply for subsidies and help securing the patent which was eventually published [173]. After the company was founded, with the help of a small subsidy from the Dutch government and many hours of consultancy, our *empathy* increased after talking to different customer segments, such as jewelers and 3D printing companies – an emerging hot application around 2013.

In a relatively short time we ultimately found a solution that works for the majority of customers we met, and developed a product still commercialized to date, the BuBble Bag [58], see Figure 3.10.

The *IF* method (Section 2.7) is partially the result of our need to convince customers about how the ultrasonic cleaning technique using bags was superior to the traditional method of using glass beakers or just suspended inside the bath – *Step 6 – calculate and decide*. A typical example of the type of calculations we used is shown in Table 3.1.

Table 3.1: The superiority of using the Cavitation Intensifying Bags (CIBs) in ultrasonic cleaning of 3D printed parts, evidenced by innovation/intensification factor (IF_t) values larger than one.

Case	Factor	Normal	CIB	d	IF
3D printed	Time [min]	8	1	1	8
	Volume [L]	100	50	1	2
				IF_t	16

In yet another innovative effort, we found that "cleaning is a dirty business," and the ways to measure when something is clean or not was (and still is) a quite difficult

---

3 The previous four steps are described in the previous section.

4 Imagine how hard it is for foreigners in some countries to find a job or how difficult it is to organize your personal life while politics and migration rules are in the way.

technological challenge. Since I had a position as postdoc in the early years of our spin-off, Bram and I combined efforts and co-edited a Special Issue on Cleaning with Bubbles which was published when I had recently joined the Science and Technology Faculty at the University of Twente as Assistant professor [98, 174]. Just imagine how challenging it was for two fresh PhD graduates to persuade authorities in the field, the editorial board of the journal and the reviewers to do their (unpaid) work (see Figure 3.10, top).

If that were not enough, the BuBble Bags were commercialized as a cleaning tool, but we did a lot more interesting things with it. If you look carefully, this container is just a scaled-up, numbered up version of the microreactor from my PhD (see Figure 3.10, bottom). With it, we have published a series of scientific articles, of which I want to focus on one to illustrate the way to apply the *process intensification approaches, SEST*, and the positive outcome of the synergy of microfluidics with ultrasound. In short, we added catalyst particles (Pd/Al$_2$O$_3$) to study and improve advanced oxidation technology for wastewater streams [145] (see Figure 3.10, bottom).

Advanced oxidation processes are used to eliminate organic contaminants from industrial waste streams as well as persistent pharmaceutical components in drinking water. We observed an enhancement due to a combination of improved mass transport, the creation of thermal gradients and the catalyst particles near the bubbles created inside the bags. The *IF* calculation can be seen in Table 3.2.

Table 3.2: Intensification factor calculations for CIBs with and without catalyst (Pd particles) [145].

Case	Factor	CIB	CIB+Pd	d	IF
CIB+catalyst	HTA concentration in 30 min [µM L^{-1}]	0.35	1.85	−1	5.29
	Reaction rate – linear behavior [µM min^{-1}]	0.012	0.096	−1	8
	Cost, metal in catalyst [euro/kg] in %	100	260	1	0.4
				IF_{total} =	16.92

Exercise X1. Can you calculate the total intensification IF_t of using the Cavitation Intensifying Bags (CIBs) in ultrasonic cleaning for jewelery parts?

Case	Factor	Normal	CIB	d	IF
Jeweler	Time [min]	10	2.5	??	??
	Volume [L]	3	0.05	??	??
				IF_t	??

Please find the answer to this question in Section A.4.

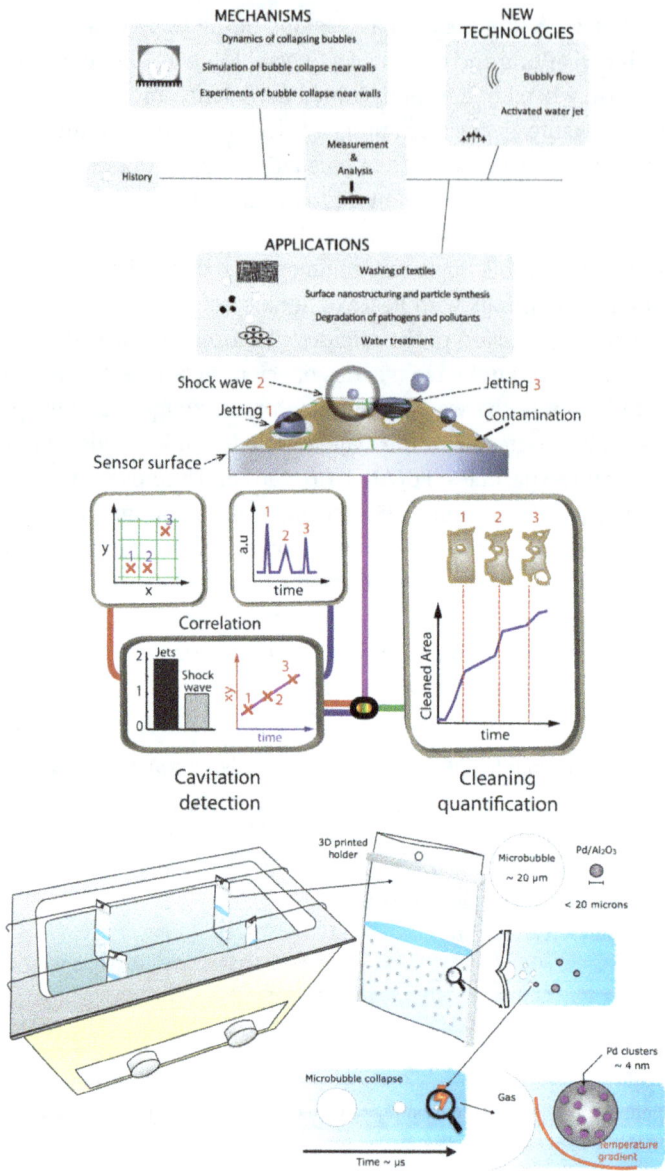

Figure 3.10: (left) Overview of the topics covered in the Special Issue: Cleaning with Bubbles, taken from [98]. (right) Ideal sensor for the simultaneous detection of cavitation and cleaning quantification. On the lower-left side of the sketch we show an ideal case where each cavitation event could be recorded, in space and time [174]. Bottom: Schematic representation of the experimental setup using Cavitation Intensifying Bags (CIBs), which in fact are the BuBble Bags. The CIBs are numbered up (to four) and placed on top of the active areas in the ultrasonic bath. Cavitation and the desired effects are enhanced by the presence of pits (indentations on the CIBs inner walls). In an example of synergy between ultrasound and catalysis, suspended catalyst particles increased the oxidation effects, the most salient feature of this study. Taken from [145], Creative Commons CC-BY license.

3.4.2 Starting another company

After joining the Faculty of Science and Technology as Assistant professor in August 2014, I had the freedom and task to expand my research focus. Around that time, one of my MSc supervisors, Surya Raghu, connected me to Ruben Ramos-García. As a result, Carla Berrospe joined my lab as visiting PhD researcher for one year, with the plan to see if it was possible to create bubbles with a rare type of lasers (continuous wave). I had already some experience making bubbles in small channels, but it was a different laser technology (pulsed) [87]. That visit led to the publication of two articles and helped me revisit work done by other colleagues at the University of Twente. Altogether, it opened up a very interesting phase in my research activities.

Almost a decade before the events I am presenting here, Detlef Lohse and his collaborators started studying how to inject without needles with pulsed lasers (see Figure 3.12, bottom). As mentioned in the previous section, page 50, there is a lot to win from collaborating and communicating effectively with different experts. The experiences of Lohse's team helped me focus on unexplored opportunities, and his team shared with me knowledge that would have been impossible to acquire in a reasonable (short) time. In particular, Claas Willem Visser became an invaluable collaborator who is featured in Section 6.2.6.

Bubbles for needle-free injections

When initiating my research to develop a needle-free injection platform technology, I realized there are many applications where it could cause impact, including cosmetics, vaccinations and tattoos. In parallel with performing experiments in the lab, my team and I took up the challenge to get out of cozy research settings and started interviewing several stakeholders in the problems caused by needles – *Step 1 – defining magnitude or problem*. The team has changed over the years, and you can see some of the protagonists of this journey in Figure 3.11.

Figure 3.11: A selection of the Team members that have worked with me since 2014.

My team's dream is making sure that the injection process is quicker than a mosquito bite, and similarly, it should not cause pain because nerve endings in the skin are not reached – *Step 2 – device first solution, milestones, reflect on limitations*. Back then, my main limitation was a lack of knowledge about previous achievements, the bioengineering and medical obstacles, and many other topics. Therefore, I set out a long-term plan to (i) acquire more information about what was possible and what had been done until then by other people and (ii) identify which problems were not only interesting from a scientific angle, but also were of high societal relevance (see Figure 3.12).

Roughly speaking, a glass microfluidic device contains a liquid that is heated by the laser (see Figure 3.13) and https://youtu.be/2AHCIgLBHu8. Within a millisecond, a bubble is created in the liquid, and the expanding bubble pushes the liquid in the form of droplets or jets at a velocity of at least 100 km/h (60 mph). These values are known to be sufficient to penetrate the skin [68] – *Step 3 – define metrics*.

I slowly managed to initiate collaborations across different institutions and disciplines. With our results, I applied for several funding opportunities, and eventually got a big research project granted (see Empathy assignment in Section 2.3.4 and its answer in Appendix A.2). In the BuBble Gun (http://www.bubble-gun.eu) project, we build on the knowledge and experiences from Step 2 and study the ways in which lasers can be used to generate tiny droplets with the right control to pierce through the outer layer of the skin, and predict the penetration depth – *Step 4 – gather information, assemble team*.

! A very good example of how to "gain" knowledge and empathy in record time can be to bring together all stakeholders you can for a short meeting. Future Under Our Skin (FUOS) is an international stakeholder consultation forum that I kickstarted in April 2019 (Event (https://www.utwente.nl/en/designlab/events/!/2019/4/152811/the-future-under-our-skin), Interviews (https://youtu.be/VpmSOrYuzFM)). It incorporates scientists, dermatologists, tattoo artists, philosophers and other experts sharing a vision on the future role that our skin has for improving the quality of life.

The most moving interaction I have had along this journey was talking with parents of small children fighting Type I diabetes and the young adults who show so much bravery in their daily activities, constantly pricking their skin and injecting insulin and glucagon for their whole lives. Understanding the actual limitations of the most advanced technology in use was only possible after talking to these people who really need a better solution. That information is not plainly written in the scientific journals or technological reports, nor can you easily get it from the doctors or medical companies that have developed it. *Empathy* enabled me to align my efforts very fast with the needs of the end-users of the technology we are developing.

Likewise, it became more evident during mass vaccinations against COVID-19, in the year 2021, how many people fear needles. The fact that intradermal injections without needles could have such a great positive impact on vaccination rates is a great source of motivation, and I empathize not only with those who fear the needles, but

Timeline needle-free injection

1866: First concept of needle-free injector, invented by H. Galante and presented by F. Beclard [44]. This injector uses a compressed spring as energy source.

1936: First patent filed on a needle-free injector (spring based), by Marshall Lockhart [45].

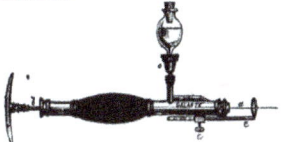

First drawing of a jet injector [44]

1956: First large scale use of a multi-use nozzle needle-free jet injectors. The 'Press-O-Jet' (spring) was used for world-wide vaccination against various diseases, such as SmallPox and Cholera [46].

1959: Introduction of the Ped-O-Jet, a foot-pedal injector developed by dr. Benenson [47].

Large scale use of the needle-free jet injector in the US Army in 1959 [62]

1966: The Med-E-Jet, the first multi-use nozzle jet injector based on compressed CO_2, is patented by Oscar Banker [48].

1996: Due to splashback and fluid-suck back, the World Health organization concludes that the multi-use nozzles jet injectors results in too large a risk of contamination, and recommends they no longer be used [49].

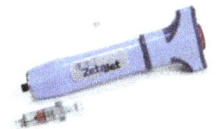

Zetajet, a commercially available spring-powered jet injector [50]

2000s: First disposable cartridge jet injectors enter the market [50], such as spring-powered Imojet [51], ZetaJet [50], LectraJet [52], Dermojet [53], PharmaJet Stratis [54] & Tropis [55], and the gas-powered BioJector2000 [50], Zeneo [56], Dose Pro [57] and Airgent/Enerjet [58].

2010s: First investigation of laser based needle-free injectors by various academic research groups [32, 33, 37, 40, 59, 60].

2020: Mirajet, the first commercialized laser-based needle free injector, acquires the CE certification for medical devices in the EU. The Mirajet is used for medical aesthetic injections [61].

Mirajet, the first commercially available laser-based jet injector [61]

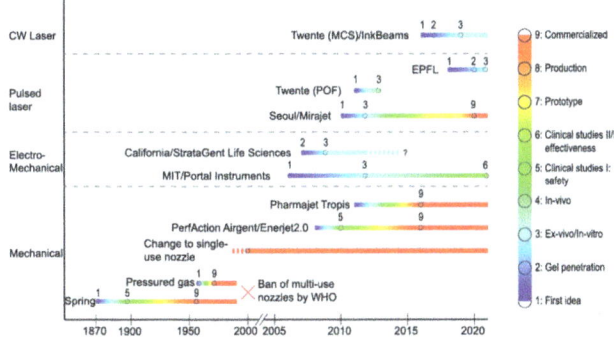

Figure 3.12: Top: Progress of needle-free jet injectors over the past 200 years. Bottom: Advancement of various technologies and research groups (with respective spin-off companies) of needle-free jet injectors, please note that the citation numbers in the figure correspond to reference [157], and not this book. The numbers relate to the ERC Technology Readiness Levels [32]. Adapted from [157], Creative Commons CC-BY license.

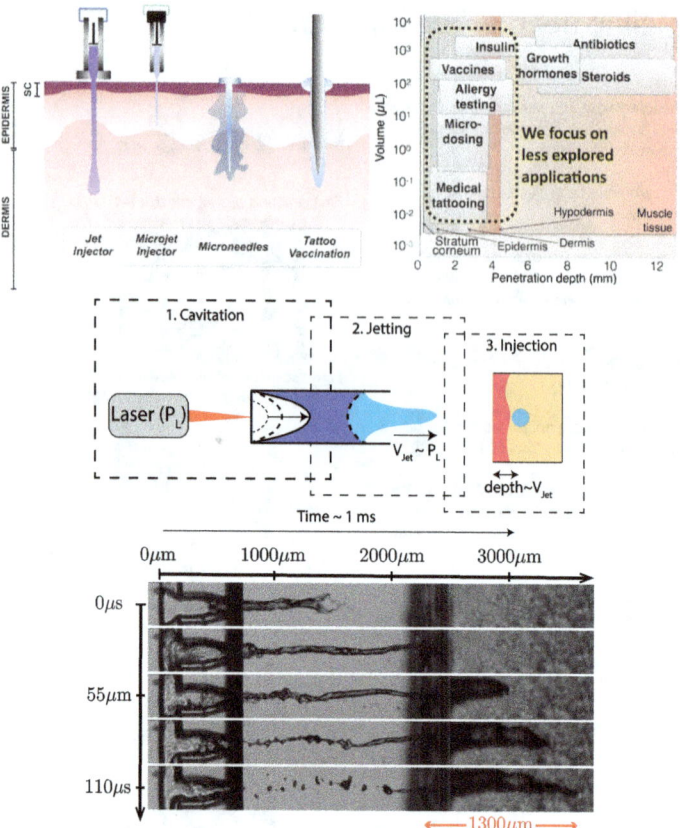

Figure 3.13: Top left: Taken from [129]. Top right: Adapted from [52]. Middle: The BuBble Gun project is split into three interdependent studies: cavitation (bubble formation), jetting and injection [157]. This image was taken from an Open Access Article licensed under a Creative Commons Attribution-Non Commercial 3.0 Unported Licence. Bottom: Image sequence showing a liquid jet penetrating a 1% (w/v) agarose hydrogel, corresponding to a microfluidic device with angle $\alpha = 14°$. For a jet speed of $v_{jet} = 48$ m/s, the jet penetrates $L_m \approx 1300$ μm. The splash-back of liquid is undesirable because the delivered volume V_D is slightly less than 100%. Taken from [142]. Please, see this video explaining the process https://youtu.be/2AHClgLBHu8.

with our environment: just think how many needles and plastic we are using with each injection and the impact it has as biotoxic waste.

For *Step 5 – Break it down*, I want now to show you how the SEST approaches discussed in the context of process intensification (Section 3.2) can be applied in this project, even when the chemical engineering component is not evident.

Structure: The velocity and shape of liquid jets can be tuned by varying the geometry of the microfluidics channels where we create the thermocavitation and jetting phenomena (see Figure 3.13) [142].

Energy: This is one of the strongest arguments to defend our innovative approach, because the idea of using lasers to create fast traveling jets was not new. In fact, I had the chance to discuss with colleagues who had tried it before, in particular with the use of a pulsed laser. We chose to use continuous wave lasers [144], which offered some advantages over the other methods [157].

Synergy: Here we are combining heat transfer phenomena, fluid dynamics and bioengineering at the microscale. The main scientific advantage of choosing microfluidics is the increased control of phenomena hard to understand or study in conventional settings (see also Section 3.3). The technological advantage is that we are riding a wave that combines the latest developments in semiconductor fabrication (diode lasers), microfabrication of tiny channels all integrated into a device for a biomedical application that gained lots of attention in the last decade: injections into skin.

Time: Whereas many predecessors were trying to inject large volumes deep into the skin, say the muscle, I chose to focus on delivering smaller volumes at shallow depths into the skin (see Figure 3.13). There are advantages from a clinical point of view, but also fewer technological hurdles [157]. Then, by using different laser sources we started studying and exploiting a new heat transfer and fluid dynamic regime. Moreover, we decided to focus on the less explored idea to deliver a larger number of jets over a larger area on the skin to avoid the limitations of conventional injectors delivering larger volumes in one spot in one shot.

The last step, *Step 6 – calculating and deciding*, is still in the process, but for now, just know that together with the University of Twente, I incorporated a company, FlowBeams, to valorize this idea in the best possible way.

You can find several videos explaining the basic idea, such as here: https://youtu.be/b7CccqKk36I (Universiteit van Nederland) and https://www.weforum.org/videos/a-dutch-scientist-has-invented-needle-free-injections (World Economic Forum).

4 Where did I read this?

Reading time ~ 50 min

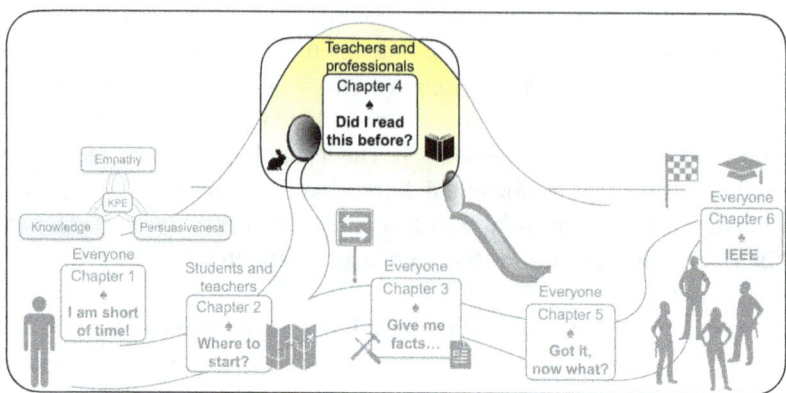

Bit: More details and knowledge sources

> "It is certainly impossible for any person who wishes to devote a portion of his time to chemical experiment, to read all the books and papers that are published in connection with his pursuit; their number is immense, and the labour of winnowing out the few experimental and theoretical truths which in many of them are embarrassed by a very large proportion of uninteresting matter, of imagination, and of error, is such, that most persons who try the experiment are quickly induced to make a selection in their reading, and thus inadvertently, at times, pass by what is really good."

Michael Faraday in 1826 [66]

In this chapter I will provide more backing and references to the ideas discussed in the previous chapters. I am almost sure there are important works that could have been cited, and unfortunately there will be omissions, which are by no means intentional. If you, kind reader, identify one author or idea that should be included, please contact me, I will continuously work to improve this book, see http://empathic-engineering.com.

4.1 Overall learning objectives of this book

I would consider this book a success if you can feel that one or more of the following learning objectives apply:[1]

---

[1] The overall goal of this book was introduced in Section 2.1, but here I provide the learning objectives that can help in case you are an instructor teaching a course or you are studying on your own.

1. You can *criticize* the definitions of engineer and entrepreneur in Sections 2.3.1 and 2.3.2.
2. You can *explain* the framework with the three ingredients of knowledge, persuasiveness and empathy (KPE) (Sections 2.3.3, 2.3.4 and 2.3.5, respectively).
3. You are able to use KPE's framework to *solve* problems of different nature.
4. You can *decide* which indicators to *use* in the *IF* method to calculate for different solutions or alternatives, *apply* it in making decisions (Section 2.7) and *assess* the Applicability of a given solution (Section 2.6).
5. You can *explain* the guide to innovate and *identify* the role of KPE in each innovation or problem solving step (Section 3.1).
6. You can *summarize* and *interpret* tips from examples of professional trajectories told by innovators about their journeys (Sections 3.4 and 6.2).
7. You can *explain* and *justify* the need for constant education to prepare you for the fast changing world we are living (Chapter 4).
8. You are able to *compile* and *use* concepts and activities needed to teach KPE (mostly this Chapter 3).
9. You can *interpret* and *criticize* a list of concepts and tips to start or keep innovating, with a focus on STEM-related activities (see Chapter 5).
10. You are properly equipped to *recognize* and *assess* the risks of certain behaviors when working alone or in teams towards a societal-relevant solution with impact on the environment and society.

Already at the end of the nineteenth century the number of scientific publications had increased to such high numbers that scientists were facing a problem that has only gotten worse in the twenty-first century. Not only do we have access to printed books and journals at the traditional libraries, but now we also have access to virtually infinite websites where any imaginable piece of work is just "a click away" – either through legitimate access or pirate websites.

As a consequence, many of the best scientists have openly and privately acknowledged that they do or do not make an attempt to "keep up with the literature," in the words of Einstein himself [104]. I would not advice you to follow this example blindly, because I am sure Einstein read, at least every now and then, between his own work activities.

Actually, research on specific topics often has peaks of interest, followed by years of limited research. Then, the topic re-emerges and part of the older knowledge is lost because researchers tend to focus on the most recent published work. Therefore, we should try to go back as much as possible – decades or even centuries if you are lucky – to prevent missing relevant information in the scientific discussion. In fact, some of my favorite papers are the oldest I have found, because they are written in less complex technical English.

My message to you is that you cannot really cover all that has been written, but you can at least structure your search and systematically "repeat" or *re-search*. As I tell my students, that is one of the jobs of a "researcher." Another important job is to write good publications, original articles, reviews and other types of documents, to make the knowledge transfer easier.

4.2 A new engineer

It never occurred to me before 2021 to define what an engineer is – see Section 2.3.1 – as I was well aware that each university or country has different requirements. As a freshman student, for me "engineers" were those who would get busy with more practical aspects, without forgetting the fundamental knowledge typically developed by graduates from physics, mathematics, chemistry, and so on.

If you randomly ask people, they will probably say that an engineer is someone who has graduated from a technical college or university. Engineers are expected to apply scientific knowledge to solve technological problems. Interestingly, engineers are often seen as good in understanding policy and organizational aspects related to technology. The best part for me is that engineers are capable of specializing in various technical and nontechnical fields, and they tend to have a "first name": electrical, electronics, mechanical, nuclear, and so on.

1. The Oxford Dictionary (https://www.oxfordlearnersdictionaries.com/definition/english/engineer_1?q=engineer) says that engineering is the activity of applying scientific knowledge to the design, building and control of machines, roads, bridges, electrical equipment, etc.:
   (a) a person whose job involves designing and building engines, machines, roads, bridges, etc.
   (b) a person who is trained to repair machines and electrical equipment.
2. In contrast, the translation from Dutch from Van Dale's dictionary (https://www.vandale.nl/gratis-woordenboek/nederlands/betekenis/ingenieur#.YbMilS8w1qs) says:
   (a) holder of an academic degree, equivalent to a master's degree, from a technical university, an agricultural university or their predecessors; abbreviated: ir.
   (b) (Netherlands) holder of a final diploma from a Technical College (*Hogere Technische School*), higher agricultural school, etc.; bachelor, abbreviated: ing.
   However, the nicest definition I have read was provided by Miguel Delcour, manager of the Royal Dutch Engineering Society. The following translation is from Dutch:
3. *Kinderuniversiteit* (Children University) (https://www.kinderuniversiteit.be/edities/hoe-maak-je-het-met-een-ingenieur/): Engineers are real makers. Crafters and inventors. They figure out how everything works, and use that knowledge to do useful (and fun!) things with it! Engineers build bridges and towers, as well as robots and machines. And they are always looking for ways to work better, more efficiently and cheaper.

The word "engineer" derives from the Latin "ingenium," which stands for intellect, talent and ingenuity. In popular settings, an engineer is someone who has a certain level of technical knowledge and specialism through training – but the word is also associated with the "engine" or "machine"; see *Technisch Werken* [23].

The story about who is an engineer is complicated when we narrow down on specific countries, where different translations or historical moments have changed its meaning. For example, in Dutch, the terms *"ingenieur"* and "engineer" are often mistakenly used interchangeably. An engineer is a University of Applied Sciences (HBO) (ing)- or university (ir)-trained technical specialist. However, since 2003 what is given is the title Bachelor of Science (BSc) at HBO level and the title Master of Science (MSc) at university level [33]. Check also the regulations in Dutch [34].

"The curse of modernity is that we are increasingly populated by a class of people who are better at explaining than understanding, or better at explaining than doing" p. 14 [164]. Then, we really need to educate people to "do more" if we want to increase access to the positive aspects of technical and economical developments, without destroying our planet in the process (see more in Section 4.3).

"We ask students to spend far too much time solving mathematical equations and far too little time thinking about the human dimensions of the problems they are trying to solve."

"Practicing engineers are called on to solve ill-posed, messy problems that do not have one correct answer that's easily found in a textbook. Students need the opportunity to confront, rather than avoid, this complexity during these crucial formative years when they learn to think like engineers."

"Engineers need to move away from a technocentric approach and adopt a sociotechnical mindset (...) [see Drones for Good: How to Bring Sociotechnical Thinking into the Classroom [35]]. By this I mean we need to start thinking about the ways in which the social and technical are always connected. These aspects should not be separated, with technical challenges going to the engineers and social challenges going to the sociologists. (...) College students today are part of a generation that has turned the tide on years of declining civic engagement. Young leaders like climate activists Greta Thunberg and Leah Thomas have begun to call a powerful older generation to account. (...) College students are increasingly ready for conversations about the ways engineers can promote a sustainable future or engage with issues of social justice."

Gordon H. Hoople in Future engineers need to understand their work's human impact [36] – Creative Commons

---

Other experts have already indicated that chemical engineering is in need of a more entrepreneurial or risk taking attitude and the importance of research commercialization must be highlighted [132]. For example, the Sustainability, Innovation, Diversity and Education (SIDE) roadmap, introduced by Javier García-Martínez [102], summarizes recommendations from various organizations and initiatives for the advancement of chemistry. These recommendations should assist in solving the interconnected challenges we are facing (see Figure 4.1).

Engineers have been using other tools or methods that can be compared with the KPE framework. It has similarities to PESTEL (see page 38), which is used to analyze the macroenvironmental factors that may have an impact on the performance of an organization [137, 76]. Likewise, I see analogies between KPE, SEST and SIDE recom-

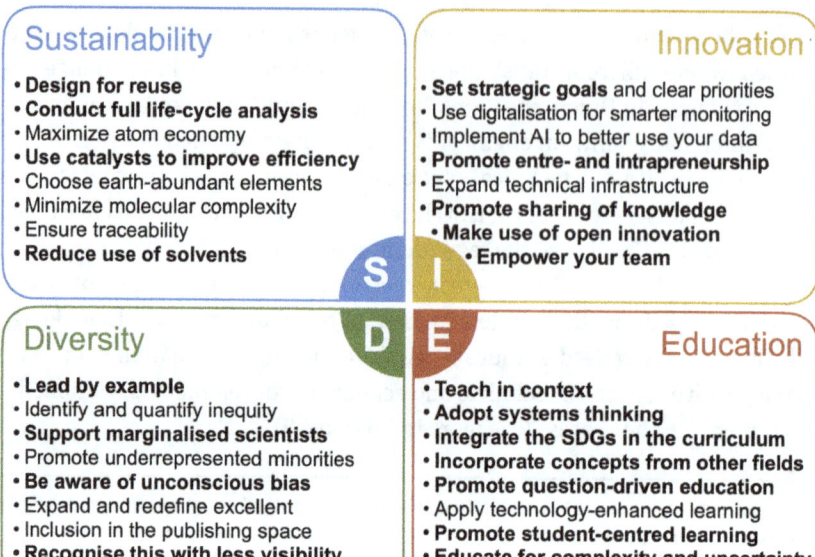

Figure 4.1: A number of recommendations meant to promote a more sustainable, innovative and inclusive chemistry. In bold I signal those tips I see are aligned with the KPE framework. Adapted from [102] with permission from John Wiley & Sons.

mendations. Highlighting all the direct connections between what I compiled in this book and SIDE is a worthy effort that could be expanded elsewhere. For now, allow me to select the following, which connect to the topics covered in this book:

Sustainability: Design for reuse, conduct full life cycle analysis, use catalysts to improve efficiency, reduce use of solvents.

Innovation: Set strategic goals and clear priorities, promote entre- and intrapreneurship, promote sharing of knowledge, make use of open innovation, empower your team.

Diversity: Lead by example, support marginalized scientists/employees, be aware of unconscious bias, recognize those with less visibility.

Education: Teach in context, adopt systems thinking, integrate SDGs in the curriculum, incorporate concepts from other fields, promote question-driven education, promote student-centered learning, educate for complexity and uncertainty.

Meanwhile, other elements related to the selection above will be expanded in the following sections and Chapter 5.

"We cannot continue to extract, emit and dispose at the levels we are doing now without compromising our climate, our planet and our own health. If we want to have a viable industry and a healthy planet, the circular economy cannot be just an aspiration but the key objective of chemistry. To achieve this goal, chemistry must evolve from being the science and industry of transformation (linear) to the science and industry of reuse (circular)."

Javier García Martínez, in [102]

The concept of planetary boundaries [37] has been instrumental to increase awareness of the limits within which humans can continue to develop and thrive, and help to think of alternatives to mitigate the risk of generating large-scale irreversible environmental changes.

4.3 Drivers for paradigm change

Scientific and technological developments have had an undeniable impact, positive and negative, on the development of our society and the environment. Few people may argue that we are experiencing fundamental transformations in the ways that society and the economy influence each other, particularly over the last two decades. A slowdown in the global economy or at least intermittence is thought to be caused or aggravated at the least by the current economical and scientific growth models: more is best, more articles published, more citations, ….

More voices are being heard, for example of the physics Nobel laureate Giorgio Parisi, warning about the blind devotion to GDP and the inability to meet targets set in the Paris Agreement. Greta Thunberg has been a vocal example of being brave enough to face the mockery of presidents of powerful countries and businesses' lobbies. I see her as evidence that younger generations do care and take action to solve things that are going wrong with our world.

Luckily, there are cases where economic growth can also help the planet, when using new "green" technologies. *Growth can be made green, even if this will require the rapid adoption of new technologies. Crucially, this will rest on the continued pursuit of research and innovation, leveraging* **human ingenuity**, *which luckily knows no bound.* Is infinite economic growth possible on a finite planet [3]? We could also think of slower growth or rebound effects: when you make something more resource-efficient, people tend to buy more of it, making efficiency gains obsolete.

Optimists like most engineers – myself included – look for ways to produce and exploit the natural resources in a more sustainable way. A large majority of inventors and also politicians and economists are sometimes complacent that this can always be supported by technological changes or innovations. But we need to make sure that any innovation should bring aspects, mostly positive, and put systems in place to identify the negative aspects not evident to the innovator and relevant stakeholders.

Take as an example the impact on the environment and other phenomena like global warming and the fact that natural resources are limited. Luckily, there is a wider percentage of the world population that has become more conscious on how we should exploit natural resources and apply technological innovations. My humble take on this is that we need to educate at all levels to achieve this needed shift.

> This is all pretty sobering, but it's even more so when one is reminded of the inevitable inefficiencies in our use of energy and the subsequent production of entropy resulting in pollution, low-grade heat and environmental damage and destruction. (...)
>
> For example, only about 20 percent of the energy in gasoline is actually used to keep a car moving. A major role of innovation is to decrease such inefficiencies by refining extant technologies, inventing new ones, or developing new ways of organizing their uses. We are witnessing increasing public and corporate consciousness of the challenges of energy consumption, waste, and inefficiency as government programs and taxation policies encourage new ways of thinking and addressing these issues. (...)
>
> It is a matter of faith that the free market system geared to open-ended growth, even when tempered by governmental intervention, stimulation, and regulation, can find a meta-stable balance between making significant profits and solving the problem of sustainability. The primary function of business, after all, is not to increase efficiencies, but to make profits. (...)
>
> A crucial element in how life has been sustained is that the energy source, namely the sun, was external, reliable, and relatively constant. (...) The truly dramatic and revolutionary departure from almost three billion years of sustainable business as usual came about in just the last two hundred years when our discovery and exploitation of the sun's energy stored underground in coal and oil heralded the beginning of the Urbanocene. Fossil fuels were, and still are, perceived like the sun itself as an almost limitless source of energy whose subsequent release sparked the Industrial Revolution.
>
> Geoffrey West in Scale [180]

The same way that scientific innovation is being accelerated in globalized markets due to our interconnected world, we need to also teach at a more global dimension. The Fourth Industrial Revolution apparently will be based on technologies such as alternative fuels, 3D printing and nanotechnology, for which not all study programs have been adapted or tailor-made.

Moreover, paired with the importance of online platforms,[2] instead of avoiding their power, for example, of social media, some scientists are seen to team up with designers, marketers and other professionals across disciplines.

I believe this is an effective way to better address the urgent needs of society, by communicating with citizens and stakeholders. All innovators should be willing to do is to "listen, read and think," and maybe show some humility – see Section 5.7.1.

It is argued that including humility in the curriculum of formal programs offered by incubators, universities and business development centers, with other values such

---

2 As I was wrapping this book, the metaverse was announced . . . .

as cooperation, collaboration and trust, can have positive implications for multicultural education and the social impact of entrepreneurial ecosystems [165, 154]. If that happens, we can induce a market-pull approach where new ideas and products are tailored to specific stakeholders, and possibly accelerate deployment into society.

> "Real problems in the real world are infinitely complex, and for any given problem, science offers only part of the picture. Climate scientists can tell us with high certainty that human activities are raising Earth's mean surface temperature, that extreme weather events will occur, and that melting ice caps will cause abrupt changes in the global climate. But it takes *time and money* to produce such certainty, and for all the doors that science even provisionally closes, others relevant to policy remain beyond closure by science alone. In the case of climate change, for example, science cannot tell us where and when disaster will strike, how to allocate resources between prevention and mitigation, which activities to target first in reducing greenhouse gases, or whom to hold responsible for protecting the poor. How should policy-makers deal with these layers of ignorance?
>
> The short answer is with humility, about both the limits of scientific knowledge and about when to stop turning to science to solve problems. Policy-makers need to focus on when it is best to look beyond science for ethical solutions. And science advisers need to admit that other sorts of analyses must also inform political decisions. Capacity-building in the face of uncertainty has to be a multidisciplinary exercise, engaging history, moral philosophy, political theory and social studies of science, in addition to the sciences themselves.
>
> We need disciplined methods to accommodate the partiality of scientific knowledge and to act under irredeemable uncertainty. Let us call these the technologies of humility. These technologies compel us to reflect on the sources of ambiguity, indeterminacy and complexity. Humility instructs us to think harder about how to reframe problems so that their ethical dimensions are brought to light, which new facts to seek and when to resist asking science for clarification. Humility directs us to alleviate known causes of people's vulnerability to harm, to pay attention to the distribution of risks and benefits, and to reflect on the social factors that promote or discourage learning.
>
> Policies based on humility might: redress inequality before finding out how the poor are hurt by climate change; value greenhouse gases differently depending on the nature of the activities that give rise to them; and uncover the sources of vulnerability in fishing communities before installing expensive tsunami detection systems."
>
> Sheila Jasanoff on Technologies of humility [111]

But now we need to talk about how we can prepare the future generations of engineers to innovate in sustainable ways.

4.4 Method to teach KPE

4.4.1 Learning from challenges

To meet the demands or requirements for educating the future generations of engineers, such as SIDE, I propose challenge-based learning (CBL), represented in Figure 4.2, as an effective way to teach and ensure better learning experiences [93]. CBL should not be confused with the concept of problem-based learning, which focuses on the acquisition of knowledge rather than on its application. Students need to be actively engaged in a challenging problem with elements of real-world context, which typically are open problems – open-ended, unstructured, interdisciplinary, etc. – without known solutions.

CBL characteristics are by design aligned with the KPE framework. For example, students go out and find solutions on their own, for which they need to talk to stakeholders, sometimes training persuasiveness to obtain information, sometimes empathizing after talking with people suffering from the problem.

In short, CBL can be split into three actions (see Figure 4.2):
1. *engage*, where students get connected to the problem and define the actual problem they will work on with the help of a tutor and interacting with peers;
2. *investigate*, where they find new knowledge and skills to understand and analyze the problem mainly during self-study; and
3. *act*, where students design, implement and evaluate a solution. The addition of challenges to learning environments results in passion, ownership and training which more readily reflects the processes in the real world.

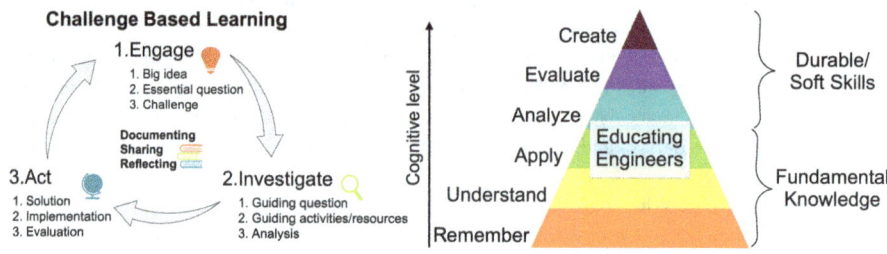

Figure 4.2: The three steps of CBL can be arranged in an ideal logical order that can be iterated several times – indicated by the curved arrows and defined as mini-investigation cycles. These cycles could correspond or be analogous to the step-by-step procedure seen in Figure 3.1, Section 3.1. The concept can be extrapolated beyond the engineering discipline, looking at the educational process, where *knowledge* and *durable or soft skills* are entwined with *cognitive levels* of increased complexity. The *KPE* framework discussed in this book can be interpreted as a simplification of the triangle (Bloom's taxonomy) in the context of innovation processes [94].

The missing – or less obvious – ingredients empathy and persuasiveness are readily incorporated into the three steps of CBL, stemming from the enhanced passion and ownership of the challenge faced. Additional learning outcomes potentially necessary for the innovative process not encompassed by KPE include traits such as courage, risk taking and the ability to cope with uncertainty.

KPE provides an educational, foundational framework to build off or to highlight and hopefully help you attain more cognitively complex attributes. In simpler words, by focusing on the different levels, you are forced to think beyond the purely technical aspects and make connections with the societal aspects of the problem.

There are analogies with the reflection cycle, ALACT: (1) action; (2) looking back on the action; (3) awareness of essential aspects; (4) creating alternative methods of action; and (5) trial, where the focus is on a personal reflection [120]. ALACT is in my view directly related to the actions of innovation and entrepreneurial goals. The ability to take action, assess past experiences to prepare for new attempts and select the most viable are a constant iterative process implicitly and explicitly followed by all innovators I know. More on the iterative aspects can be found in Sections 2.9, 2.6 and 3.1 and Figures 2.12 and 3.2. The reflective actions are expanded in Section 5.6.

An important element that needs to accompany CBL and KPE is seen at the center of the left side of Figure 4.2: documentation, sharing and reflection of each of these steps – more on this in Sections 5.2 and 5.6. Without it, the actual process can be lost or will be difficult to use in the future. The evidence or reasoning behind every decision can help "persuade" others or allow for reporting at least when and what influenced each aspect.

It is not surprising that, though the CBL concept was developed as a pedagogical tool, you can find parallels with KPE. In a way, the self-learning sought by CBL is what I believe corresponds to self-discovery of solutions to problems, but perhaps in a more realistic setting, say, outside the classroom.

To fully understand the cognitive processes at work within the three steps of CBL, please take a look at established educational frameworks such as Bloom's taxonomy [55], see right side of Figure 4.2. Borrowing from these pedagogical sciences which are applicable to the process of innovation, together with a group of experts, we proposed a visualization of how to combine Bloom's taxonomy with the "chemical engineering toolbox" required in process intensification [94].

This visualization is not comprehensive, as it does not include transferrable skills such as a critical mindset, interdisciplinary collaboration, communication and information literacy, which are listed among the "21st Century Skills" [46]; see also A Comprehensive Guide to 21st Century Skills [24].

The *KPE* framework discussed in this book is instead a simplified version of the triangle representation of educational challenges [55]. This simplicity allows it to be adaptable to other disciplines and more broadly applicable.

4.4.2 Teaching knowledge

You will probably feel uncomfortable reading the name of this section. I agree with you if that is the case, because we normally do not stop to think about the meaning of it. We intrinsically "know," or believe we do, when we are in command of a notion or idea.

It is not my intention to enter a philosophical discourse, or define even more terms and concepts than I dared to posit in Chapter 2. That would mean talking about discriminating what is wisdom, whether we should teach skills, the value of practice and a long list of interesting topics, where knowledge is not only technical, but can also include network, personal traits, know-how and the ability to gather information and to cope with uncertainty [56].[3] I repeat my fair statement, you learn a lot by teaching!

In short, I do not believe that there is an optimal way to teach that fits all students. Likewise, I have not met a good teacher that keeps teaching the same topic with the same tools for too long. This is because, even when you get lucky and find a method to teach a specific discipline or concept, with time, so many things will change in society, science and technology that you will probably have to start all over. See the Knowledge assignment on page 10. You probably can complement the reading of what follows next.

You may be interested in a kit for innovating or ways to learn how it is possible to "think in new ways." It is a good idea to think of the work ahead as systematizing what you learn, and arguably you will need to communicate, to be able to innovate; but more importantly "converse with yourself" [148]. More about this topic can be seen in the context of self-reflection in Section 5.6.

> The history of human thought would make it seem that there is difficulty in thinking of an idea even when all the facts are on the table. Making the cross-connection requires a certain daring. It must, for any cross-connection that does not require daring is performed at once by many and develops not as a "new idea," but as a mere "corollary of an old idea." Isaac Asimov Asks, "How Do People Get New Ideas?" A 1959 Essay by Isaac Asimov on Creativity
>
> [25]

As you probably will find in several parts of this book, I believe humor and cultural references do help in understanding and communicating serious engineering work. In fact, there are different ways to break the traditional storytelling, or push against naturalization: when, for example, cultural or other phenomena are gradually seen as natural.

One useful technique I like is "*defamiliarization*," in which you frame usual things in a different way [13]. Such approach can help identify "taken-for-granted" technologies, accounting for ethnography and the role of information and communication technologies [51]. In simpler words, you can speak about *redefining the problem* or being prepared to address new problems.

---

[3] The topic of uncertainty in business settings has been laid out specifically for physicists in Davide Iannuzzi's book [110].

Ultimately, applying this tool can reveal the how to use new or old technologies in unexpected, but hopefully useful ways. For example, humor uses dissonance or strangeness related to a preconditioned way of thinking [169]. After the incongruity or strangeness is resolved, we typically laugh – or not if it is a bad joke. But even bad jokes can help us reinterpret a notion or think differently at the least. More about this follows in Section 5.4.1.

Allow myself to take a short detour to talk about a pedagogical aspect. Currently most education experts agree that "constructivism" is a very useful theory that helps explain how learning happens [118, 70]. It is based on the idea that the learner constructs the knowledge scaffold and its meaning by himself. For example, students build their own study track, defining their meaning, and are able to gain new knowledge. In many cases, students are more familiar with direct instruction, which can be part of good education, for example helping the learner construct meaning.

In my view, an ideal constructivism situation is when the student chooses among courses or elements in a course and is guided by what the teacher defines as the objectives' success criteria and overall process of a course. The teacher then provides feedback and checks if the learning objectives are met. The student, in turn, provides an evaluation about the whole experience which should help the teacher improve the course or activities. As with many interpersonal activities, very often "one size cannot fit all preferences." My experience is that sometimes a given course edition can have great success with the majority of students, and in others not, depending on the level of guidance or freedom the students demand.

Direct instruction is often good for transmitting basic knowledge and to explain well-established procedures. In contrast, anyone who has seen open-ended problems solved by student teams knows that it can get quite chaotic and that it requires lots of time. Moreover, often teachers and students feel that many things could have been done better. This is the challenge I face when teaching in subgroups during my courses: how much individual or group instruction or coaching is needed? My answer will not come as a surprise to you: I believe the world needs learners that will be able to solve challenging problems and are able to work in teams, with the teacher's help as a coach while solving such problems.

Arguably, knowledge about something, with sufficient practice time, can be perceived as or becomes a skill. Marike ter Maat, a teaching expert at the University of Twente, proposed to me an analogy with teaching: "You need knowledge about several theories, then you start teaching and try to refine it. However, you always keep the class and the individual students in mind, and teach in the way they like and learn."

4.4.3 Teaching persuasiveness

I think of the challenge of teaching persuasiveness as a "durable skill," which increases with the degree of "softness." What I mean is that conventional, gained knowledge and skills can be demonstrated via traditional tests: you pass a driving license exam, you successfully complete a simulation training with a software program, you are able to discuss or write about a theoretical or abstract concept, etc.

I am not aware of study programs where STEM students are trained to persuade explicitly. Following from Section 2.3.4, business schools are probably the place most people would look when interested in gaining other soft skills related to persuasiveness, such as decision making skills, time management, etc.

Even when my suggested Persuasiveness assignment on page 12 has not yet been rigorously tested from a pedagogical perspective, I have had many discussions with colleagues about my answer. As a result, we have concluded that the simplest way to teach persuasiveness is to provide cases – from the real world and imaginary – that can simulate situations where it is needed. Thus, this book is loaded with several examples.

You can see, for example, K as knowledge taken from a book; P is practice with cases from real life in the university-academic setting or during internships at industry locations, where the student needs to persuade as part of the assignment, for example, to defend the choice of using given software or equipment; E can be gained during an internship or bachelor assignment inserted in a real research environment, company or university, by talking to more experienced professionals and if possible with other stakeholders, such as the ultimate beneficiaries of a given solution.

In project-based learning it is not necessary to involve stakeholders. However, in CBL activities you are expected to talk to several stakeholders for their feedback. For example, Dalton could talk to his brother and clarify the source of their problem (page 36, Section 2.10). In most of my courses I design a process that simulates a real-life scenario that can be found in academic settings or the business world. I include three basic elements: (1) in-depth knowledge of an emerging field of research and current industrial relevance, along with (2) work in teams and (3) an appropriate professionalization component. If you look carefully, this idea I had in 2014 is the seed of what KPE became in 2021.

The first element is basically gaining information that should become knowledge valid for the final test or teamwork, and there is a mix of individual work and work in teams. The teams are presented then with a problem.

In most cases, I am able to offer the students to work on real-life – open, i.e., there are no known answers to the questions – problems provided by my industrial or academic network – labeled as "case owners." Sometimes, some of my research team members play the "case owner" roles as well. Then, the students go out on their own to complement or expand their knowledge and propose a solution.

The (proposed) solution is written in a report that is peer-reviewed among the groups. The revised report is later discussed in a presentation in front of the "case owners," as if trying to convince them or persuade them that their solution makes sense. I give the students tools, such as the *IF* method (Section 2.7), so that they can compare and demonstrate superiority of ideas – this book is actually the scaffold of all my courses.

In parallel, I meet with the students and show during lectures or individual team meetings how I would typically approach other problems with analogies to their project. It is important and difficult at the same time to give ideas without giving the answers or solutions to their problems, but it is a feasible challenge. As a teacher, I keep on asking questions to the students, persuading them to answer all these questions, and discuss, individually or in groups, why they should persuade (metacognitive approach).

My idea of a good persuasive story is based on three main aspects:
1. simplicity of the solution and how good its explanation is;
2. proof of validation of the need by a given stakeholder; and
3. the credibility of the proposed action plan that leads to the execution of the problem's solution.

The *first* is possible to assess by having other teams of students perform a peer review round. Project-led education has been identified as an efficient active learning method, instrumental to learning to "understand" [140]. Having different projects but the same starting point (knowledge acquisition), these students are in the best position to give feedback and assist me in helping each team to polish their idea.

The *second* can be in the form of an interview made by the student to a stakeholder or statements quoted from a reliable source, such as an interview in the public media or found in scientific publications.

The *third* is definitely the hardest, because my courses are not that long to allow for executing their proposed solution. Sometimes we have the chance, which depends on existing lab installations in my group or at the case owners, to do experiments or build a prototype. But, if you have worked on a technology development project, you will know that plans are broken almost from day 1 of the project execution.

The preparation to face milestones or expectations not reached is another explicit focus in my teaching and a source of frustration for many students (see Figure 5.2). That is actually a necessary experience to face the real world.

A last note on how to improve persuasiveness has to do with how you convince.

> For the sake of persons of different types of mind, scientific truth should be presented in different forms and should be regarded as equally scientific whether it appears in the robust form and vivid colouring of a physical illustration or in the tenuity and paleness of a symbolic expression.
>
> James C. Clarck, in [133]

What I want to transmit here is that you need to find the best way to communicate – with students or stakeholders in real-life innovation projects – because not everyone understands or sees a concept or idea the same way.

4.4.4 Teaching empathy

Teaching empathy has been identified as a vital aspect, not only for human–human interactions, e. g., health professionals [147, 113], but also for machine–human interactions [124]. Similarly, design thinkers learn to observe, interview and develop empathetic insights with the aim to find human-centered ways of solving problems [60].

I am currently gathering information and reaching out to people with more experience in teaching, with the aim of building an equivalent way to teach empathy in STEM educational programs [79]. From these interactions, I share some ways a teacher or teaching team can transfer durable skills. Here are some examples of how to teach and how these activities lead to an increased understanding of empathy in the students:

- Teachers are aware of their own behavior and become an example to the students.
- Teachers provide the students with explicit elements about KPE, with abstract concepts and sufficient case studies, as I provide in this book.
- Teachers implement KPE in the whole curriculum, in a learning line of academic technical and durable skills. Ideally, these notions can be included in several activities, and not necessarily in a single course focused on durable skills.

Whatever the chosen activities, it is about implicitly and explicitly working with KPE and focusing on individual levels and the group level (teaching team).

The challenges and controversies of teaching empathy have been discussed in ethics, social science, history, medical and legal training [105, 113, 181], and are also impacting fields where computers are trained to communicate with humans [112, 124]. Attempts to conceptually and empirically distinguish among various types of emotional responses and measure their relations to empathy have been made through physiological indexes, facial indexes and self-report indexes [134, 78].

In the case of humans, it seems that empathy can be developed as emotional intelligence through providing examples and role models – see the assignment on page 15. With the growth of civilization and cognitive abilities, we gained the ability to familiarize ourselves with diverse and distant cultural groups, beyond our families [63]. As most who have studied or lived outside their country of birth, international education broadens the capacity to empathize. This appears to hinge on a person directly engaging with another culture to actively learn, rather than passive interaction such as through reading about the culture.

This idea is supported by reports evidencing that to learn in a meaningful manner, students must be actively engaged in the learning process [114, 85]. It is not only a matter of engagement, but also of serious reflection on what the students have learned and experienced. This is also consistent with my understanding that you must advance to higher levels of cognitive complexity throughout the KPE framework.

You can find several revised entrepreneurship education policies and academic programs [139] to assess the influence of entrepreneurial training on how well it prepares young people to be entrepreneurs. More on the need for an entrepreneurial mindset in STEM is given in Section 4.5.

In my courses, I regularly invite different stakeholders, from founders or CEOs of companies to professors from other disciplines (see Figure 4.3). The presence of other professionals or experts in the classroom helps broadening the view of the students about their future. They are triggered to imagine different roles for themselves in the near future, because it can help clarifying existing conceptions about a given role (CEO, CTO or lab technician). I believe this can increase your understanding and consequently empathy when deciding to start your own or join an existing organization.

Figure 4.3: Bram Verhaagen visits my classroom in 2015, as CEO of BuBclean at the moment. He and other visitors shared experiences of what it is to be an entrepreneur. Bram also appears in Figure 7.1. The students are seen here working in groups and "playing" a competition game to assess their knowledge after listening to the visitors.

4.5 The need for teaching empathy and persuasiveness

This book is not the appropriate place to elaborate much on the philosophical or deeper reasons of why teaching the right type of empathy or persuasiveness is important (see Sections 2.3.4 and 2.3.5). But, roughly speaking, some people seem to have an innate ability and others are capable of acquiring it along their life path.

What I want to make sure is that those who do not have it, or are not convinced of its usefulness, can see my point on its value. Let us now focus on how we can use it to solve problems.

Empathy can be utilized in two main ways:
1. to identify a problem affecting others that represents an opportunity for solution creation; and
2. to create a compelling case, to convince others of either the importance of a problem or the impact of a potential solution.

With respect to (1), an entrepreneur or problem solver must understand what has been done before and avoid what is popularly known as "reinventing the wheel." It will help finding out how the current method or product is underperforming for those currently using it and devising a way to compensate that insufficiency.

The development of this understanding typically requires interaction with other people. When humans interact, (2) there are several emotions or transactional activities that are relevant in most professional fields. In particular for effective communication, professionals interfacing with customers or healthcare professionals interacting with patients must connect to the experience of the customer or patient.

As rightly observed by some of my collaborators, there is an apparent overlap with point (2) and persuasiveness as the ability to communicate a case which stems from empathy.

However, there are boundaries for this type of connection. For example, some study programs train healthcare professionals to distinguish between sympathy and empathy, teaching that one must be empathetic and not sympathetic, as sympathy will hinder appropriate diagnosis. Too much sympathy can also negatively affect emotionally the professional providing for the customer or patient.

I hope you can agree with me in the fact that sympathy is not appropriate or sufficient for a STEM professional – see the Rolling Stones quote on page 15. No matter if you are an innovating engineer or not, sympathy does not necessarily provide an accurate picture of how a problem impacts other people. To succeed in valorizing an innovative idea, cognitive empathy – defined in Section 2.3.5 – may not be sufficient either, since the detached tone associated with it does not help in the important action of persuasion.

Moreover, to succeed in having the much needed paradigm shift towards sustainability in so many areas of human activity, we need to ensure the future generations of innovators master an empathic approach to work – see also Sections 4.2 and 4.3. However, a vast majority of professional education programs or literature makes little or no effort to consider the role of empathy. When empathy is addressed, it is typically empathy of instructors for their students that is stressed, rather than how instructors can impart empathy through their teaching activities [166, 49].

In particular, STEM careers are traditionally focused on providing as much information as possible, training in using some tools or procedures and hopefully knowledge. Unfortunately, in contrast, clear indications or practical guides to transfer inventions from the university environment to the market are rarely given. In this note, see Section 4.6 for examples for guides for knowledge transfer to society. An empathy-centered program would emphasize making a real-world impact. Kamp recently argued that human literacy is about empathy, communication and the ability to connect people by putting relationships at the heart of the work [116].

"It's not just a matter of translating jargon into plain language. As Kathleen Hall Jamieson at the University of Pennsylvania stated in a recent article [15], the key is getting the public to realize that science is a work in progress, an honorably self-correcting endeavor carried out in good faith.

Moreover, scientists need to have some understanding of their audience to improve the chance of a true dialog. They may need to learn to listen and 'read the room,' and prepare different approaches for different audiences. (…)

Rather than focusing on the data showing why smoking is dangerous, the campaign revolves around personal stories of patients who have suffered smoking-related illnesses that individuals can relate to. Application of this principle – which meets folks where they are rather than burying them with data – led to remarkable progress in smoking cessation. Understanding this is a high-level skill that requires expertise."

It is also the case that not every scientist wants to take time away from research to be a voice for science. Traits that make a good researcher, such as concentration on details and laser focus on a problem, don't often carry over to the public stage. Most scientists prefer *to persuade* by performing meticulous, credible work.

H. Holden Thorp [168]

Traditionally, there is a general conception that scientific work happens in places where ideally everyone carries a lab coat and all materials and equipment are available to the researchers. Even when I have seen some of these labs, the reality in most innovation workplaces is quite different. In the last few years I happily saw the publication of books written by academicians who carry inspiring messages and inform us of the challenges of doing science. These books depict innovation and entrepreneurial activities, sometimes implicit, and in others the main focus or final aim is to help scientists becoming founders or be more entrepreneurial professionals. I humbly believe this book will help to bring more good examples from what happens in the labs to society.

4.6 How others teach entrepreneurship, E_2

A course on entrepreneurship for engineering typically will contain the following learning objectives [80, 42]:
- learn what it takes to become a technopreneur;
- understand the various ways of identifying opportunities;
- acquire the knowledge to conduct market research and provide evidence for the viability of the business idea;
- develop a viable business proposition;
- understand the dynamics of new venture development and team building;
- recognize the value of entrepreneurial ecosystems and put them to good use;

- translate the business idea into marketing and financial plans;
- present and pitch your business plan to various audiences, such as investors, customers and partners; and
- depending on the school, you may be able to test a market hypothesis with an MVP (minimal viable prototype/product).

 It may be interesting for you to check theoretical knowledge and practical skills required to transform innovative ideas into commercially viable businesses, which is called "technopreneurship." The book Technology Entrepreneurship: Bringing Innovation to the Marketplace [16] gives some tools and frameworks to manage, commercialize and market technological innovations.

In the view of Kamp [116], there will be more emphasis on engineering as a social activity and empathy with customers and colleagues, together with concepts of autonomy, agency, self-efficacy and emotional intelligence to function in the social environment. More elements can be found in Section 4.4.

4.7 Four recent books

In what follows, I will share some elements from books that I have selected for a simple reason: I know the authors and I have had the chance to discuss KPE and how our approaches complement each other with them.

4.7.1 Book 1. Tropical Empathy

Chances are that you may be facing challenges to perform high-quality research in a less developed country or in a rich country with limited resources for research or educational activities. It gets worse when you have to work against political restrictions. If you want to know, for example, how it is to do research under "tropical" conditions with a very creative attitude, this author has several other publications in that line [45].

> In contrast to disciplines like medicine or biology – where individuals such as Carlos J. Finlay and Felipe Poey became world-recognized figures of Cuban science more than a century ago – physics has only a modest tradition in the country. The most prominent Cuban physicist in the first half of the 20th century was Manuel F. Gran – an inspiring teacher and an enormously cultured man who published detailed and rigorous physics textbooks, but almost no original research.
>
> The situation was more or less the same until the late 1950s, when a social transformation catalyzed a number of radical measures: from nationalizations of foreign companies, to the Reform of Higher Education in 1962. One of its enthusiasts would say: "It was a time of little control, so one could swiftly do a lot of good things!" One of those "good things" was to immediately force scientific research into Cuban universities.
>
> In October 1960, the US imposed an "embargo" that has been evolving right up to the present day. It makes it extremely difficult to purchase US-made scientific equipment in Cuba. However, collaboration with the former socialist block – especially the Soviet Union – as well as individual

scientists from Europe, Latin America, and even the US, has helped the development of Cuban physics since the early 1960s.

Soviet electronic equipment was heavier and probably noisier than its US counterparts, but it worked. Thanks to it, Cuban physicists were able to achieve many things in a couple of decades, including fabricating solar cells, and designing original experiments to be performed in microgravity conditions during a joint Cuban-Soviet space flight in 1980, to cite two examples. (...)

My strategy to survive as an experimental physicist in "high tropicality conditions" was to violate the boundaries of "safe science", invading zones where I was not a specialist, looking further afield for new phenomena, seeing scientific instruments in everyday objects, attacking and retreating from serendipitous findings like a guerrilla. Facing challenges became a natural part of my way of doing science, and I started to feel an immense sensation of freedom. Now I'm addicted to it. (...)

When I look down at a pile of junk garbage, I see scientific setups. If the reader suspects that some of the experiments described in previous chapters are just garbage, in the present chapter he (or she) is absolutely certain to be exposed to garbage science.

Ernesto Altshuler, in [45]

4.7.2 Book 2. Entrepreneurial physicists

This is an attempt to help physicists in academic settings to deploy good ideas to market. Providing know-how of great relevance to the technology transfer process, you will find key ingredients, theories and models to assist in becoming entrepreneurs and generate value for society – of relevance not only to physicists I must say. The book includes discussions of effectuation theory, internal resource analysis, external landscape analysis, value capture, the lean start-up method, business canvases and financial projections. There is attention to "often neglected" aspects such as trust, communication and persuasion. Not surprisingly, I agree with the attention given to persuasion [110].

> Throughout your entrepreneurial adventure, you will need to persuade people to join your group, help you, follow certain directions, buy your products, and so on. (...) It is thus worth asking: what is it that makes people say 'yes'? To answer this question, there is no better way than looking at the work of Robert Cialdini, who was able to summarize the basic traits of persuasion in only six principles [64] (...):
> 1. Liking: we are more likely to say 'yes' to someone we like than to someone we do not like.
> 2. Reciprocity: we are more likely to say 'yes' to someone who has done something for us than someone who did not.
> 3. Social proof: we are more likely to say 'yes' to an offer that has already received the concensus of a community we are part of than to an offer that is not popular in our surroundings.
> 4. Consistency: we are more likely to say 'yes' to an offer that aligns well with what we have already manifestly supported than to an offer that would not fit well with our previous decisions.
> 5. Authority: we are more likely to say 'yes' to someone who is considered as an authority in the field than to someone who is not recognized as an expert.
> 6. Scarcity: we are more likely to say 'yes' to a rare offer than to an offer that is readily available to anyone who wants it.

(...) as for many models that pertain to the area of psychology, Cialdini's six principles of persuasion should not be taken as a checklist to be used to force people taking decisions in our favor. (...) First of all, if you do see that someone repetitively refuses your offers, even if you think that the offer was well balanced, you may want to check which of the six principles you are violating. Maybe you sound too arrogant (violating the like principle), or maybe the person in front of you interprets that offer of yours as a request for a favor, and she/he has no reason why they should do that (lack of reciprocity). Or you are proposing something that is either not yet approved by the community of potential users (lack of social proof) or against a choice that that person has made a couple of months before (lack of consistency). Or the person in front of you believes that you do not really know what you are talking about (lack of authority) or that you are proposing that offer to the entire world, and that there is nothing really special in it (lack of scarcity). Knowing what is not going well may help you clarify any misunderstanding, reposition your offer (if possible), and get more attention from the people you are talking to.

Davide Iannuzzi, in [110]

4.7.3 Book 3. Hunch Engineers

When you read that innovating is for doers and that it is possible to build ideas with a "hunch," that is in my view a persuasive statement to dive deeper. It is in a way a democratization of the invention or innovating business – socializing the entrepreneurial myth, perhaps? The main message I got was a reinforcing experience that when a problem is found, you learn not only by getting it solved, but also by being wrong. I aligned totally with the statement that "no thing is new" and that we typically see innovation as the result of trial-and-error of innovators, thus the value of being "productively wrong" [148].

"It is easy for aspiring entrepreneurs to characterize their ideas using their best understanding of those concepts in the abstract. It is more difficult for them to realize that whatever they end up with may walk and quack like a startup but not yet be a startup. (...): Incipient entrepreneurs confuse their initial guess of a destination with an actual plan of action. Unfortunately, it's easy to fall in love with the craft that goes into articulating a concept using precise technical management terms while losing sight of the job ahead. It's the same as burying yourself in technical jargon from whatever field you've been working in. Both are excellent examples of over-engineering – something every engineer is strongly encouraged to avoid.
(...) fields of inquiry – particularly those concerned with engineering, with high technology, with science, with tinkering, and more generally with the synthesis of new ideas – have yet to offer viable strategies for you to engage in entrepreneurship and innovation that are compatible with that world view. In a way, entrepreneurship and innovation emerged first as a scientific and management field, but they still lack an experimental and engineering footing.
Chemistry went through this same process before chemical engineering emerged. A symptom of this lack is that we see more people concerned with idea selection than we see people concerned with actually producing innovations. The real impact of this shortcoming is that more and more aspiring entrepreneurs and innovators focus on new consumer products and on leveraging reasonably commoditized technologies (e. g., the Web and apps).
Meanwhile, fewer pay attention to opportunities in more complex systems and new technologies or use either to conceive entirely new categories of activity. They also fail to address

meaningfully how to scale up their ideas until they become viable business concepts to which they could then apply what they have learned (or can learn) about management and entrepreneurship. (…) I liken innovating to learning (…) that occurs while you are engaged in a very general kind of problem solving, with no guarantee that you will come up with a solution. (…)

Innovating by making real-world problems tangible offers you an alternative to the many innovation recipes that have emerged from product design, product marketing, lean manufacturing, and technology readiness – the recipes my students have in mind when they ask the "contaminated" questions. To get started, all those recipes seem to require a well-formed idea about a product, a user base, or an organization – that is, they require that a large part of your innovating be fixed before you can even begin. (…) Most of these recipes seem to take a "good idea" as a given, and hinge on convincing others that it is, indeed, good (…) That is quite the opposite of learning. This book is about learning."

Luis Perez-Breva, in [148]

4.7.4 Book 4. Chemistry Entrepreneurs

As I was finishing some last details of my book, Chemistry Entrepreneurship, edited by Javier García-Martínez and Kunhao Li, was published [103]. I am still reading it and I can tell you that it is a great book for you if you want to start and grow a new company in the chemistry sector. Several authors provide information about different aspects concerning the creation of a new chemical enterprise, to create products from their research. It shares elements of being a scientist and entrepreneur at the same time.

García-Martínez has argued before that the most effective way to adapt chemistry research and industry to the current worldwide situation is to reimagine chemistry education. "If we keep teaching chemistry like in the second industrial revolution, we will produce excellent chemists for a world that no longer exists" [102]. I would add that empathy is needed in all ways to this aim, teaching the students to be able to understand the challenges faced by their teachers and also companies and other institutions to join efforts.

"Innovation, not only in products, but also in new business models and management, has proved to be the most effective way to adapt to a rapidly changing demand, regulation, and supply chains. Innovation starts by setting clear strategic goals and priorities and choosing leaders who nurture and value talent, vision, and boldness."

Javier García Martínez, in [102]

4.8 Solving changing problems

A small "eye-opener" I like to give to my students is that large transnational companies have been shedding their research and development groups for decades. In several cases, these labs or divisions become start-ups on their own. The spun-out team of researchers forming these companies can help sharpening the motivation of academics and other institutes, as the competition for subsidies or research funds and the stakes increase.

This situation has had a tremendous impact worldwide on research as a whole and the engineering disciplines, particularly in the chemical industry. In some countries, there has been a trend of governmental financing towards private investment or public–private partnerships. If we then realize the power of social media, it is no surprise to see how researchers, funding agencies and companies have initiated countless social campaigns as a mitigation strategy.

Everyone wants to show how relevant is whatever they are doing, because subsidies and funding schemes are not so abundant. Researchers, myself included, are increasingly active in social/professional media to showcase their research; see Section 5.4 and the attention note on page 100.

For their part, companies make sure to highlight how their products meet nonconventional demands, such as being vegan, CO_2 neutral, etc. Moreover, nontraditional ways to fund research for innovative applications have been funded through crowdsourcing schemes.

A crucial discussion point when preparing education curricula is: *how many of these tools and how much specific knowledge do our students really need?* Well, there is no simple answer to that question, and unsurprisingly I do not have the best menu for all interested parts. However, I take this opportunity to transfer some bits of my own acquired "wisdom."

To educate engineers to be more entrepreneurial, "change" must be the focus since students are in the process of becoming something they previously were not [79]. *This focus needs to happen as an integral part of the whole study program, i. e., over various semesters/modules, and not isolated as one (elective) course.*

Logically, the results in the classroom do not always match what happens after they graduate [179]. Therefore, engineers must be educated in attaining the right attitudes, motives and failure acceptance capability, willing to start over again when needed – more on measures of success or failure is given in Section 5.11.

This is consistent with the emphasis given to agility and adaptability regarding innovation on page 83. In pedagogical terms, this can be seen as "*transfer*": a student learns how to calculate whether a wooden beam will fracture under a given load during the education years. But in the real world, there is a need to make similar calculations for plastic beams, so there is a transfer of knowledge from wood to plastic.

There is also a possibility that engineers acquire these skills without formal training or have them intrinsically due to personality and unique experiences. It is argued that entrepreneurship education and training directly correlates with positive entrepreneurial outcomes [150]. In contrast with other curricula, entrepreneurship students may require wide but not so deep knowledge on a specific discipline, as long as they are willing to learn in the process or incorporate other team members that do possess the required deep knowledge.

Engineering education experts can learn from these experiences. For example, engineering students normally go through higher education with specific performance goals. There are typically (high) expectations to earn good grades, whereas the focus should be on having as goal the mastering of the content and durable skills [107].

A reasonable way to help teaching how continuous change is tackled in reality is to use iterations or loops to clarify a problem or innovation. For example, the four-step customer development process gives insight into what makes some start-ups successful [54].

In the context of innovation, progressing from a hunch looks like a cycle where:

> "You can generalize problem solving to innovating: Bring the problem to a resource-friendly scale, make it tangible and scale up successive demonstrations of the problem. (...)
> 1. At the start, you a have hunch.
> 2. You give your hunch the structure of a problem. You have to postulate what makes your hunch solvable, recognizable and verifiable.
> 3. You reformulate your problem at a smaller scale so you can start working on it.
> 4. You sketch the smaller-scale problem with a prototype of the problem's form and function, much as you would draw a figure to aid in problem solving.
> 5. You can now describe your prototype as a small-scale illustration of a problem.
> 6. You restate the problem."
>
> Taken from [148]

This cyclic depiction is also present in the iterative suggestions in Section 3.1 and Figures 2.12, 3.2 and 5.1.

Here follows another example of iterations or virtuous cycles:

> Thermodynamics both grew out of the industrial revolution, which provided physicists with examples (such as the steam engine) of heat at work, inspiring them to investigate just what was going on in those machines, and fed back into the industrial revolution, as an improved scientific understanding what was going on made it possible to design and build more efficient machines.
>
> John R. Gribbin, science writer [104]

If you want another example in a systematic and organized way to solve problems, you could check the theory of inventive problem solving (TRIZ/TIPS) [38]. It stems from the work of Genrich Altshuller (1926–1998), beginning in 1946. It started from studying patterns of invention in the patent literature.

The main goal of TRIZ/TIPS is obtaining an algorithmic approach to the invention of new systems that helps also to refine existing ones. This line of work has identified three main findings:
1. Problems and solutions are repeated across industries and sciences.
2. Patterns of technical evolution are also repeated across industries and sciences.
3. The innovations used scientific effects outside the field in which they were developed.

Without attempting to draw direct links with what we are covering in this book, just see the effect of avoiding "tunnel vision" and the implicit relevance of collaboration across different fields. Point 2 resonates with the now widely accepted need to work in multidisciplinary teams, and I see similarities in point 3 with defamiliarization (see Section 4.4.2 and [13]).

The patterns of inventive solutions identified over the years and the first analytical TRIZ tools are summarized in 40 inventive principles [26] and a contradiction matrix, where a conceptual solution is found in a diagram by defining the contradiction which needs to be resolved.

Improving the conceptual solution will get you closer to the ultimate ideal result. Altshuler went further with an algorithm of inventive problem solving (ARIZ), which is a list of about 85 step-by-step procedures to solve contradictions.

(...) to maintain open-ended growth (...) the new dynamic must still be driven by the positive feedback forces of social interaction responsible for innovation, and for wealth and knowledge creation. Such an "intervention" is none other than what is usually referred to as an innovation. (...) The entire sequence is continually repeated, thereby pushing potential collapse as far into the future as the creativity, inventiveness, and resourcefulness of human beings allow. This can be restated as a sort of "theorem": to sustain open-ended growth in light of resource limitation requires continuous cycles of paradigm-shifting innovations.

On the grand scale the discoveries of iron, steam, coal, computation, and, most recently, digital information technology are among the major innovations that have fueled our continued growth and expansion. Indeed, the litany of such discoveries is a testament to our extraordinary ingenuity.

There's yet another major catch (...) to sustain continuous growth the time between successive innovations has to get shorter and shorter. Thus paradigm-shifting discoveries, adaptations, and innovations must occur at an *increasingly accelerated pace*. Not only does the general pace of life inevitably quicken, but we must innovate at a faster and faster rate! (...) explaining the contraction of socioeconomic time and the increasing pace of life.[4] (...)

Unfortunately, there is no established quantitative "science of innovation" and therefore no universally agreed-upon criteria or data relating directly to major innovations and paradigm shifts, let alone to finite time singularities. So in order to confront theory with data we have to rely on informal studies and a certain degree of intuition.

This situation may well change as innovation becomes an increasingly active area of investigation, with researchers beginning to grapple with questions such as what is innovation, how do we measure it, how does it happen, and how can it be facilitated?"

Geoffrey West in [180]

---

[4] See Section 2.2, where I explain the boundary conditions within which this book was written.

5 Got it! Now what?

Reading time ~ 60 min

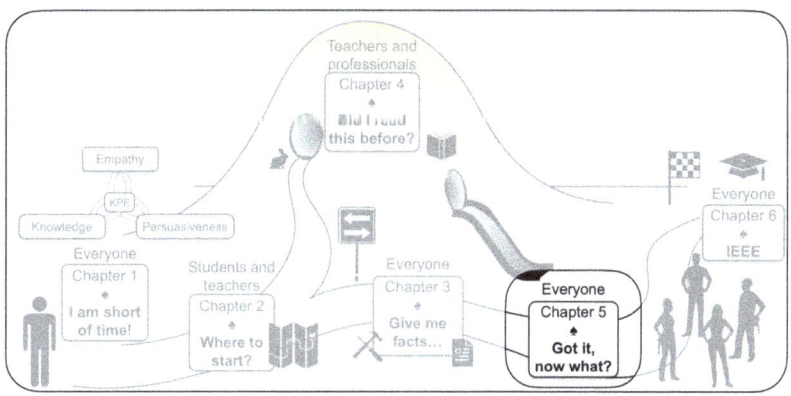

Bit: I give you personal tips

> "Success is not final, failure is not fatal.
> It is the courage to continue that counts."
>
> Winston Churchill, British statesman

In this chapter, I want to expand on some additional considerations and advice.

We will go into a more "personal take" on a few topics I believe can help you further in your own career path development.

Actually, all that you have in this book is perhaps the least resource-intensive method to innovate I can imagine.

Arguably, if you have some brains and the will to put it to work after reading these chapters, you not necessarily need expensive equipment or lengthy training periods to get started.

Let us start this chapter by summarizing what we have covered so far:
1. A proposed definition of engineers and entrepreneurs (Sections 2.3.1 and 2.3.2).
2. A framework with three ingredients (KPE) that can help you solving problems: knowledge (Section 2.3.3), persuasiveness (Section 2.3.4) and empathy (Section 2.3.5).
3. You can now use a tool to assist you in making decisions (Section 2.7). The *IF* method is a way for decision makers to arrive at a numerical approximation when deciding whether to continue expending resources during an innovative process.

I hope it will help when used in combination with the Applicability proposition introduced in Section 2.6.
4. You can follow an innovation guide (Section 3.1). These six steps embody the principles of the KPE framework. Please do not think of them as a checklist of processes which innovators must rigidly follow. They should be interpreted instead as a demonstration of the links between knowledge, persuasiveness, empathy and the actions undertaken in the pursuit of innovation.
5. Loads of additional reasons and tools that clarify the need for constant education to prepare you for the fast changing world we are living in (Chapter 4).
6. After this Chapter, you will get examples from my own professional trajectory, and from other innovators' journeys through case studies from the real world (Section 6.2).

5.1 Take-home message

This book tries to make the point that a more explicit inclusion of persuasiveness and empathy education in technical fields is needed to enable effective and sustainable innovation. The KPE framework may help define common ground that facilitates the exchange of best practices and results between professions which do not generally emphasize empathy, such as engineering, and others which do to some degree, e. g., medical and legal professions.

Broadly speaking, engineering may be typically viewed as a problem-solving discipline, where the technical aspects are either emphasized or the sole focus. My collaborators agree that STEM professionals-in-training – that is, you – need to be taught that the output of such efforts should include a dimension of empathy [96].

Likewise, you should be given tools to develop both empathic and persuasiveness abilities in parallel with tools to develop knowledge. As presented in Section 4.4, the principles of CBL are a prime candidate testing ground for integrating these new ways of teaching. The following tips and discussions are my humble effort to help you at a more practical and personal level.

5.2 KPE³ tips

Going back to my definition of engineer (E_3) and entrepreneur (E_2) (see Sections 2.3.1 and 2.3.2), I want to combine it with the KPE ingredients, such that

$$\text{KPE} \times E_3 \times E_2 = \text{KPE}^3. \tag{5.1}$$

This new acronym, KPE^3, will serve the purpose of putting together a graphical representation that accompanies six tips as shown in Figure 5.1.

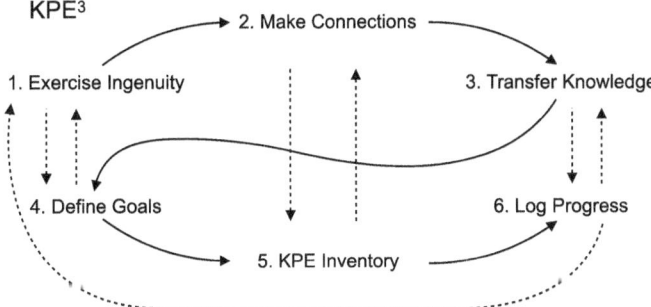

Figure 5.1: The six steps that summarize the "coaching" session that this Chapter 5 is: KPE^3. The dashed arrows represent a reciprocal supporting role or interdependence between the pointed steps. This is a continuous process, even if it looks like it ends at Step 6 (a dashed arrow suggests going back to Step 1). It is like a game of Snakes and Ladders that never ends.

1. **Exercise your ingenuity:** This will allow to widen your "tunnel vision" and be more creative, and it will force you to look beyond the specific area of the problem you are solving, for example, by using concepts or tools from other disciplines not directly related to yours. If you see how architects or designers approach a project and the way they predict the interaction with surroundings, it could teach you something about how to design a room with complex equipment you are designing or building.
2. **Keep making connections:** These connections are meant as between you and other people, but also associations with different contexts. Besides the point made before, this will allow you to acquire new knowledge, ideally from different sources and people. You can think of starting a new hobby or taking a beginners course in a new discipline in which you have no past experience. Perhaps taking a trip to a remote area as a volunteer will enrich your experiences and increase your empathy for those who have a totally different daily routine.[1]
3. **Never stop transferring knowledge:** Teaching others, you learn a lot about the topic or information or knowledge you are transferring. Seeing how others assimilate the information and interact with a prototype can help you refine your solution. Observing how others apply the learning experience in their own lives is an enriching event that will always lead to positive outcomes. I take this as a responsibility or duty; after having the opportunity to learn from others, the least we can do is pass it on to others that may do good things with it.

To strengthen the first three steps, you could try and implement in loops the following three tips as implied in Figures 2.12, 3.2, 4.2 and 5.1.

---

[1] Note that retirement age seems to increase every couple of years. Thus, there is a higher chance that you will be working way into your seventies. How are you planning to stay active?

4. **Define goals:** Clearly write down your personal and professional goals. It can be a list of ideas or a schematic drawing; there is no need to make it too fancy. Doing it will help you to communicate them to your colleagues, supervisors and family. If you do so, you will achieve an alignment between your dreams or interests and the support you may get from your network.
5. **KPE inventory:** Assess what levels of KPE you possess for the tasks or projects you have in your current state and how these can help you reaching your goals (defined in the previous step). Identify how to compensate or top up to the level required to make progress, and also anticipate what you may need for your next steps. As a risk mitigation plan, try to think of at least two scenarios: one where you reach your goal and one where the goal is partially achieved, or not at all.
6. **Log progress:** Keep good track of all your actions and when they happened – yes, also your hobbies and nonprofessional activities. You may want to have a projected timeline where you put different milestones associated to a given period (in a month or year from now). This last step is very important because it will allow you to define how far you are from your defined goals and to see if any adjustments are needed with KPE.[2]

The following sections are meant to help you executing the tips given above.

5.3 Less academic take on persuasion and empathy

While I was writing this book I was more alert to the topics we have discussed so far. I was happy to find several news items and nonscientific literature focusing on the importance of persuasiveness and empathy. This section is a nonexhaustive compilation of these miscellaneous items. We start by what I think is a good way to understand what persuasiveness and empathy are: let us look at their opposites.

5.3.1 Persuasiveness vs propaganda

Communication designed to manipulate thought or behavior, good old propaganda, is the opposite of persuasion [4]. We should not diverge much from our STEM-driven path, but it is important to be aware that unfortunately, propaganda is not only related to the traditional political parties – even when politics do influence research funding or the ethics of science and technology. Leaders of any field can distort the ability of their followers to think rationally. We must not confuse persuasion with manipulation, and engineers need to always consider what ends are served with their inventions.

---

[2] You may want to check ALACT in Figure 4.2.

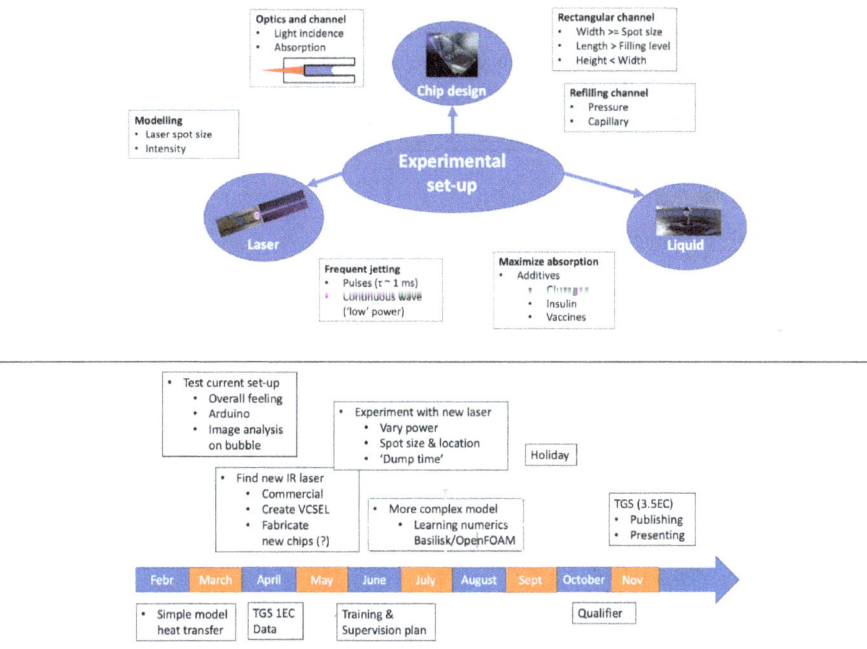

Figure 5.2: Top: Mindmap belonging to the project of a PhD student working on needle-free injections (see Section 3.4.2). Bottom: Corresponding timeline divided into months. Notice that we consider "Holiday" as it is important to keep a good life–work balance.

In the very scarce "free" moments I had writing this book, I got to learn more about the biography of Albert Einstein, particularly thanks to the series Genius (https://en.wikipedia.org/wiki/Genius_(American_TV_series)), season 1. There are two particularly renowned scientists that I want to mention to show how due to propaganda, brilliant minds can be enlisted to work in controversial topics or discriminate against fellow humans: a dangerous road to follow.

Fritz Haber was a Nobel Prize winner who is famous because of both the destructive and the constructive side of his work. He found how to synthesize ammonia, which can be used to manufacture nitrogen fertilizer, but it can also be used as nitrate-based explosives and gas warfare. The fertilizer quest was very important at a time of food scarcity in Europe. I do not know all the details, but it seems that he and Albert Einstein fell out of friendship after Haber supervised the use of chemical warfare by the German army during World War I. Each of them had different ideas of what a scientist should do in times of war. You can find more on the website of American Scientist (https://www.americanscientist.org/article/life-and-death-chemistry) and on Wikipedia (https://en.wikipedia.org/wiki/Fritz_Haber).

Philipp Lenard was another Nobel Prize winner, who scapegoated Einstein as a Jew who was contaminating Arian physics. "There is no better example than Lenard to show that a Nobel Prize is no guarantee of wisdom, humanity or greatness of any sort, and that, strange as it may seem, the award can occasionally provoke feelings of inadequacy." See How 2 Pro-Nazi Nobelists Attacked Einstein's "Jewish Science" (https://www.scientificamerican.com/article/how-2-pro-nazi-nobelists-attacked-einstein-s-jewish-science-excerpt1/) and Wikipedia (https://en.wikipedia.org/wiki/Philipp_Lenard).

As professionals, we need to protect the innocent and be careful of red flags such as poor evidence or conflicts of interest. These aspects are particularly sensible, for example, when talking about vaccination hesitancy or abortion rights, but this book is not the place to expand further on this. My aim is to raise awareness and ask you to pay attention to what your efforts are contributing to.

On persuasion, there is a last interesting angle I wanted to share, tied to advocacy or engagement discussed by Perez-Breva [148].

At a high level, advocacy, inquiry, and verification all support the same objective: A very specific "something" is somehow preventing your path to scale-up, and you need to overcome it in order to continue. You do not have the means to do so, but "someone" else does. You discover what the "something" is that you need as you make your problem tangible; you need to engage and persuade the "someone" who has it to give that "something" to you and see what he or she may want in return. The "something" might be information, insight, contact, a part, *knowledge*, or money. The "someone" can be any of the people you encounter in your innovating. That engagement involves the following:
- You talk about the problem you've made tangible, and why it matters (...).
- You state what is needed to progress. (...), this may come from how you verify the problem is solved or from what any solution needs to accomplish.
- You show a tangible demonstration of your problem – that is, your functioning innovation prototype that illustrates the larger reality you aspire to conquer.
- You ask for the resources you need to take your next steps – that is, the next set of parts and insights that emerge from your interrogation of your prototype.
- You demonstrate as tangibly and sensibly as you can how what you've asked for will translate into the progress you said was needed – that is, you explain your next steps (...).
- All along, you listen in order to find out how you are wrong. (...) You don't accept comments at face value; rather, you argue with facts and drive the conversation to what in the opinion of those you converse with would make it all right.

5.3.2 Empathy vs apathy

Maybe I'm too emotional
Your apathy is like a wound in salt
Maybe I'm too emotional
Or maybe you never cared at all
Well, good for you, you look happy and healthy
Not me, if you ever cared to ask

♪ ♩ Olivia Rodrigo (American singer-songwriter and actress) – good 4 u, 2021.

5.3 Less academic take on persuasion and empathy

The shortest tip I have for you to work on increasing or gaining empathy is just to communicate with other people.[3] You can show concern or interest with the ultimate users or beneficiaries of the solution to the problem you are solving – see a painful example of lack of transparency in Section 5.8. Talk to them, pose as many questions you can imagine, and do not forget to write down the answers. The main advantage is that you will be able to balance the expectations for you and the people you are working with or for whom you are solving a problem.[4]

Such an empathic approach will help you identify different angles of a single problem or redefine various problems, assigning priorities or levels of importance for each stakeholder. If you want, see it as "tinkering" with the information. Tinkering in this way will help you aligning the information you already have and gain the knowledge that will help prototyping a solution or several solutions.

I found very recently this website elaborating on five exercises to help you build more empathy [43].
You should combine this part with Sections 4.5 and 4.4.4.
1. *Strengthen your internal resources* by thinking about something you are struggling with and how it makes you feel.
    - Imagine a friend with that same problem and how you would respond to them. This exercise can indicate the difference between how we treat important people in our lives – hopefully with patience, generosity and forgiveness – and the way we treat ourselves – possibly with blame, harshness and self-criticism. Building self-compassion is important to increase our capacity for empathy.
2. *Feeling spent? Spend kindness on others* by doing something small for someone else, an altruistic action, such as reaching out to someone in need.
    - Interestingly, this is advised particularly when you are overloaded or stressed. In the words of Dr Zaki: "Students are happily surprised to find that when they give to others, they don't end up depleting themselves (...) Happiness and well-being are not a zero-sum situation."
3. *Disagree without debating* when having a conversation with someone who does not agree with you. Share how you formed your opinion first, and listen afterwards how that person reached theirs.
    - This is a difficult task because people tend to debate on the obvious differences, and you must be cautious to spot intentional denigration or discrimination in any way. This exercise stems from "deep canvassing," and is used as a 10–15-minute, two-way *persuasive* conversation. You should not focus on persuading the other, but try to learn how to disagree without disliking the other person or labeling them as enemy. "Empathy does not mean condoning – but it can mean understanding" according to Dr Zaki.
4. *Use technology to connect, not just to click and comment*; though not designed to build empathy itself it should help in making online platforms kinder and more humane.

---

[3] An important aspect you may want to consider is empathy towards other cultures (past and present) as well as animals.

[4] Developing a needle-free injection technology in times of mass vaccinations has put a lot of pressure on me, receiving almost every week one or more requests to collaborate, answering questions to the media, etc.

- Ideally, digital interactions are supposed to help us to connect with others, for example, by having conversations with clear text messages or actually calling by voice, rather than sending confusing emoticons.
5. *Praise empathy in others* and contribute to an overall culture of kindness.
    - This is as important as complimenting other people on their (traditional) achievements, such as passing an exam or getting a promotion. The value of this exercise is to make people around us more aware of what empathy is about, for example, by recognizing when someone in a team has helped others achieve their goals.

These exercises I trust you can do on your own, thus, there are no answers provided in this book.

The importance of empathy is highlighted because it helps us see beyond stereotypes, prejudices or biases. According to Dr Jamil Zaki [183], it means that it can help us understand people from other walks of life, races, generations or ideologies not like our own. Moreover, it seems to benefit the empathizing person because of reduced stress and satisfaction with their lives and personal and professional relationships. As a sort of warning, Dr Zaki stated that "empathy can (...) sometimes give us tunnel vision, in wanting to help some people over others."

As I confessed earlier, I am a great fan of science fiction and comics. Perhaps one of the least famous members of the Guardians of the Galaxy is Mantis, who "is an empath with the ability to sense other people's feelings and alter them" (see Figure 5.3). I wonder if she is not that popular among fans because she has no big muscles or special superpower, besides empathy, a "soft skill." In contrast, the much more popular Black Widow is "one of the most talented spies and assassins in the entire world."

Perhaps, fans of the Marvel Universe do not empathize with Mantis due to her isolated life. Mantis was naive and had little experience with social interactions. This is a sort of cautionary tale about what happens when you do not develop social or "durable skills." If you come across as an apathetic person, it will be very hard to get people to work with you, or your employability chances can decrease. I believe in aiming for a balance; a Mantis with a bit of the skills of the Black Widow may be an even more popular superhero.

5.4 Some advice and biases

To start solving problems it should not really matter where you come from. However, there are several barriers we must overcome to achieve success. To just name a few, there is bias faced by people who are not from wealthy countries, or based on gender, age, skin color and many other factors. You can find lots of reports or ideas to work under adverse conditions (see the note on page 80 of [45] and [170]).

In sum, we may not be able to change where we were born and where we studied, and some events are definitely beyond our control, such as armed conflicts or natural disasters. But we can always "sharpen our tools," or in other words, gain new knowl-

Figure 5.3: Contrast between Mantis (right) and Black Widow (left); the more empathic one is not as popular as the highly skilled assassin. Drawing based on characters from Marvel Entertainment.

edge and be aware of possible biases influencing us and those we are interacting with. There is a recent attempt to help visualizing every single cognitive bias [5]. Over the years, these mental mistakes or biases[5] have been studied with the aim to understand how we think and act and to identify irrational shortcuts causing several problems in entrepreneurship, investing or management.[6]

For example, it is common to extrapolate information from the wrong sources, pursuing the confirmation of pre-existing beliefs and "failing to remember events the way they actually happened!" One that I just want to summarize here is the *self-attribution bias*: "An entrepreneur overly attributes his company's success to himself, rather than other factors (team, luck, industry trends). When things go bad, he blames these external factors for derailing his progress." You can safely replace the word "entrepreneur" with "engineer" and not much would change.

---

5 There are 188 known confirmation biases identified, ... and already people are concerned about artificial intelligence bias; see Battling bias in AI, BBC Story Works [39].

6 Not all behavioral scientists agree with this visualization because there is overlap and some frameworks are contradictory. For us, "engineers," let it be a visual aid to increase our awareness.

If we then accept that it is human and common to be mistaken, we must focus on improving the way we work and study. Look for good ways to train your persuasiveness, ensuring it does not come across as plain manipulation, and try to empathize with people facing problems. This is precisely why I have collected the following pointers and framed them within my proposed KPE framework.

5.4.1 Off- and online pointers

We have discussed in different parts of this book that science, innovation and technological developments are ultimately a social activity, or simply said: done by humans. It should not come as a surprise that it all boils down to communication. Therefore, the tips and experiences I will be sharing here could probably be available in literature or other sources you would not readily relate to STEM disciplines.

The way scientists and inventors have been communicating over the centuries has changed continuously, for example, using paper and ink or computer-aided design (CAD) tools. But there is a common factor across the generations, and that is the dialogue between the protagonists – letters exchanged or other written forms where ideas, concepts and prototypes were laid out and the way other people interpret them.

In the past, a book would be written and the author would hope that someone could be able to understand the concepts after reading the book. Nowadays, we have videos and multimedia that can tell us almost in real-time how different stakeholders are interacting with a given innovation or scientific output. But the fact that people are downloading a given document is no guarantee that it will be useful for them. How to make sure our ideas or inventions are making an impact in society is a very good question that escapes the scope of this book – but I am very interested in knowing all about it.

Let us get back to the main objective of this section. If you start reading a patent or an article from a professional journal, you will not always understand at first sight what is being said. You may need to have prior knowledge or some sort of training to find where each piece of information is given.

Also very important, you should be sure to understand how credible the source is. The same happens when you compare a publication or written correspondence from the eighteenth or nineteenth century with the latest published work you can find online. I do not think I have the ultimate or right answer, but I suspect it all depends on how well informed you are and if you can find where the active discussions take place or where the latest developments are being made public.

In short, perhaps some advice I give you in the upcoming sections may lose relevance in a few years, or even months; but there are three counter-arguments:
1. These elements are relatively universal, i. e., they do not age – with perhaps some changes in the role of technology, e. g., messages normally arrive faster via email than via conventional mail.

2. What I provide here has been enriched with input from students, experienced professionals and even nonscientist colleagues of mine: for me that is a validation of the value for now and the near future
3. If these elements run out of fashion too soon, it might be a good reason to have new editions of this book to update them over the years.[7]

5.5 Communication etiquette

Nowadays the exchange of ideas and information mostly happens online, and this will probably continue to increase. Moreover, the COVID-19 pandemic during the last couple of years made it even more difficult to organize in-person events. Thus, I have collected a nonexhaustive list of tips that can help you keep a log of your actions, prepare emails and letters and participate in conferences and professional meetings, both in person and online.

Meeting new people
- Learn from them, and let them know about your ideas, goals and plans. You may need to use different ways to convey your message: simple graphics, short videos, text, equations, etc. (see page 101 for an example on aphantasia).
- Try to always do your "homework" before approaching someone, particularly when you are trying to gain the attention or are in need of their support. Check their *curriculum vitae* (CV; what they have done professionally), recent online activity, publications and interviews given to the media.
- This background check is important when you want to align your interests with theirs, particularly if there is something you could provide to an ongoing project led by the person you are corresponding with.
- Can you arrange a "warm introduction"? This is when you have a contact person in common.[8] See also Section 5.5.1 on writing tips.

Starting or growing collaborations
- The ability to clearly state what you want can help you get the job of your dreams, join a new team or obtain a promotion, investment for your project, etc.
- Align your efforts by validating your KPE; see Section 5.2 for tips.
- Recruit or clearly signal you want to be recruited for any cause worth your time and that of your interlocutor.

---

[7] I am sure my editor and publishing house will love this idea.
[8] My visit at MIT in 2017 was only possible with the support of several contacts who helped me get to the recipient of my emails, who typically has an overloaded inbox.

Logging your own activities
- Having a good method to keep track of your actions is probably a must. This can be in the form of a diary or notebook where you leave notes for yourself – this works online or with good old paper.
- Self-reflect or discuss with yourself if a given action led to the result you expected, describing how you think the levels of KPE have increased since the last time you did the steps or followed the steps given above.
- A good log helps you keeping your CV updated, which can then help other people when approaching you (see item *Meeting new people* above and Section 5.6).

5.5.1 Writing tips

To communicate effectively and be persuasive, you most likely will have to talk or write about what is happening in your head. I think it is too difficult to include in this book advice for your verbal communication and to pitch persuasive presentations [75]. Instead, I want to tell you how I write these days to communicate with other people.

The author of the book *Digital Body Language: How to Build Trust and Connection, No Matter the Distance* [72] said in an interview that people with connectional intelligence know "to never confuse brevity with clarity, that *reading carefully is the new listening, and writing clearly is the new empathy*" [74].

Connectional intelligence prioritizes deep, quality interactions, which contradicts commonly used measures of virtual success, such as numbers of connections, likes or online meetings. She has committed "to building a movement of knowledge and training (…) the skills of the new post-pandemic era," e. g., understanding when a meeting can be a call or when to look directly at the camera during an online event.

Whenever writing a document longer than a few paragraphs, it helps me sometimes to read each paragraph backwards. I just recently found that this is a way of *defamiliarization* – see Section 4.4.2 – used in different fields, particularly in art and literature.

I want to share a warm childhood memory from Fraggle Rock – a 1980s TV show created by Jim Henson – where Muppet-like creatures live in a sort of world parallel to ours. A recent post in a blog [13] where defamiliarization was discussed made me think of the most entrepreneurial example I had while growing up:[9] Uncle Traveling Matt. Every chapter, Matt sent a postcard – yes, a piece of paper with handwritten

---

9 There is no time to enter this interesting discussion, but just imagine that I grew up in a socialist country, where the concept of entrepreneur is not precisely what most people imagine: starting up a company to just make money.

notes! – where he would update Gobo Fraggle, the main protagonist, on his travels and experiences in the world of humans, or as he called it "Outer Space," full of Silly Creatures (see Figure 5.4).

If you have a chance, try to watch a couple of fragments where you will see that Uncle Matt not only shows us the adventurous and curious character of many entrepreneurs, but also teaches us how humans' regular activities can be analyzed from a different angle – enhancing our *ingenuity*. When we notice something that is common for the majority, there is a big chance you can innovate or solve a problem in a novel way – remember the glass of carbonated water from Section 2.3.3?

Figure 5.4: Uncle Traveling Matt sending a postcard. Courtesy of ™ & ©2022 The Jim Henson Company. All Rights Reserved.

5.6 Models and dialogues with yourself

In this book I make use of a few acronyms (KPE, SIDE, IF, TRIZ, ALACT, etc.). They help me to memorize concepts described in a given method or guide that are too complex if you would try to recall them word-by-word. When you are logging your own actions or communicating within a team where everyone knows about specific acronyms, that is fine. You should avoid jargon or acronyms that are not known to the people you are communicating with. This brings me to another common feature in the professional literature, the (often over-)use of models – and corresponding acronyms.[10]

> It's not just students who are pleading for a ready-made innovation model to be handed over. (...) This is why there is hardly any management book published anymore without a hip canvas model or another type of model. (...) We can do fine without all those innovation models. When we rely on the creativity of our brains, we can go a long way. (...) Because the creativity of our brains quickly causes us to jump from left to right, we tend to reach pretty quickly for a model. And models can also be very helpful. (...)
> We first need to have a good idea of the problem that the model is intended to help us solve. (...) there is always a model to be found somewhere that can help you out of the quagmire.
> (...) It is our creative brains that make new discoveries and figure out how to do things faster, easier or in a more satisfying way. It is not the models that do that. Innovation means entering into a conversation with each other, exploring the space outside the box together, inspiring each other with new knowledge and insights, (...) don't forget that ultimately it is a group of people who are crafting that innovation together.
>
> Eveline van Zeeland in Without dialogue, every innovation model is worthless [40].[11]

As mentioned above in Logging activities, you should always keep an updated CV, no matter how early in your career you are. Perez-Breva said that "The benefits of documentation will be most easily understood by future-you – the real expert at what you are doing and fully equipped with hindsight" [148]. Even as a student, though it may sound silly, you have to start somewhere; see also its use for reflecting and planning your strategies in Section 5.6.

What may seem simple or irrelevant for you now could actually tilt the balance in your favor when recruiters assess your trajectory, for example, if you have volunteered or taken a project away from home or anything that stands out or is exceptional – see Section 6.2.5. Since most of the times you have very limited information of who your competitors are, you better make sure your story is consistent with who you claim you have been and where you are aiming at.

I am not saying that everyone needs one, but starting a blog was an idea that was suggested by a seasoned professor I was consulting when preparing for applying for research funds; see the Persuasiveness assignment in Section 2.3.4 and the answer in Section A.2. I thought that at my junior faculty position level that was a bit pretentious, and my record was not as impressive as those of the other scientists I knew who kept an online CV or blog up-to-date.
 I took the advice, and eventually learned that indeed, sometimes there are people really "checking" what I put on my news section (https://david-fernandez-rivas.com/index.php/category/news/). More importantly, it helps reviewers of my work to check what I have been doing in the past. But it can also help students when deciding if they want to follow an assignment or course that I am giving.

---

10 I had to use some of them in the book, sorry.
11 See also "Contaminated questions" on page 82.

Nowadays you can make such cool pages with embedded videos, interactive items, ... it can be a "rabbit hole" at times. Nevertheless, I try to keep it up-to-date, and sometimes it helps me updating my CV!

If you feel writing your own blog is not something you are ready for now, it may be good to read other people's blogs. In that way, you can learn about useful things that would not be in your regular field of work, avoiding "tunnel vision." For example, I recently learned about aphantasia [6], the inability to visualize that affects 2 % of people. Steve Blank's blog is probably the last place where you would think you may learn about it, because he is more known in the field of entrepreneurship, as creator of the customer development method that launched the lean start-up movement.

The fact that not everyone can close their eyes and imagine something they have not seen yet can make you think twice – *empathize* – before you take for granted that people can imagine things that easily. It connects to a previous *tip* where I suggested to prepare good graphics in parallel to accompanying text, since not everyone "sees the same way."

Another blog example from which I learn about different aspects in a less traditional way is via Barking Up the Wrong Tree [7]. Eric Barker presents science-based answers given by experts to gain insight on how to be "awesome at life." One of the ideas he has been presenting over the last decade is WOOP – yes, another acronym – which stands for Wish Outcome Obstacle Plan and helps to assess if your dreams or plans are realistic.

You can do this WOOP exercise with almost anything you can think of. The validation that it is working is that if you get energized, the plan is probably a good one. If you continuously feel de-energized, the wish may not be realistic and you may want to try smaller changes: from dreaming to doing.

W. Your dream, a new job or any other vague wish you have **O.** This needs to be a specific goal, such as become your own boss or join a given company. **O.** The difficult part: What is blocking your way? You may not know where to apply for the job or who is interviewing. **P.** You set up a plan with actionable items, e. g., to talk to your network, someone you know who may have the right connections to get you to who you need to see, or whatever you may need.

Eric also argues that money is an easy measure of success, but trading time for money is perhaps not the best way to look at it – see Section 5.11 for more on this. Instead, four areas of focus have been proposed to track feeling of success, and another acronym emerged before us: **W**inning, **I**nfluencing, **N**eeded, **E**njoying (WINE); cheers! WINE corresponds to the following motivations or feelings:

Achievement: feel like you are winning, by setting achievable goals – could be short and small celebrations;

Legacy: feel like you are influencing others in a positive way, helping others; as a teacher or entrepreneur you can do this with students and employees;

Significance: feel that you are needed or valuable for those around you; being patient, loving, being reliable; and

Happiness: you need to enjoy daily what you have in your life, what you have achieved.

If you are following all these tips so far, you may see the many similarities among them: sense of duty or serving others, focusing on your own happiness, and so on.

But focusing on all these tips at the same time is close to impossible. In this respect, it would be very difficult for me to recommend all the great books or blogs that I have read about managing your time or projects, and how to achieve that "almost impossible" focus.

I typically forget most details of any book I read, or idea I get while having a conversation. This is the case after a few weeks unless I practiced any of the tips or suggestions. But, I still remember a passage I read when defining what to do after finishing my PhD. David Allen proposes six levels, called "Horizons of Focus" in Getting things done (GTD) [44], that allow you to put some perspective to the many goals you may have:

1. life;
2. long-term visions;
3. 1–2 year goals;
4. areas of focus and accountability;
5. current projects;
6. current actions.

When we are in the middle of a difficult period, such as a transition from student into graduate life, or moments with serious personal issues, it is difficult to focus on big picture goals. Having this division of levels can help you gain some control from the low level, or the day-to-day tasks that can help to free up mental space and move to the upper levels.

In GTD you will find the need to have a Trusted System to help you cope with all the activities or projects you are involved in – you see the tip to log all your actions again, as in KPE[3]. The system has three main phases:

Capturing Get it out of your head – keeping records of all that pops up in your mind that could have a relation to any of your ongoing projects (compare with CBL; Figure 4.2).

Processing Regularly go through your Captured elements, checking your notes or drawings, and define if you need to act upon it now (if it can be executed in less than 2 minutes) or at a later stage.

Defining the actionable items that would take you more than 2 minutes should help you either placing a reminder for a future moment when you should do it or making a Next actions list (urgent) or a Follow-up list (when you need feedback or input from other people).

Reviewing The ultimate goal is to try to keep your mind free of all the small items and deadlines, if you manage to execute the steps with some discipline. That discipline is achieved in the Reviewing phase, which is the Horizon of Focus listed above. There is an analogy between this step and the *Log progress* step from page 89.

Most of the tips in this section you can practice on your own, and you could spare a bit of attention to the value of reflecting about your own actions. This is important not only for your planning and setting goals – your strategy and *Log actions* step introduced earlier – but can help you interact better with the people surrounding you. Finding the right level in reflection can enhance your professional development. For example, another "funny" name for you to remember or make you cry when chopping it is the *onion model* (see Figure 5.5) [120].

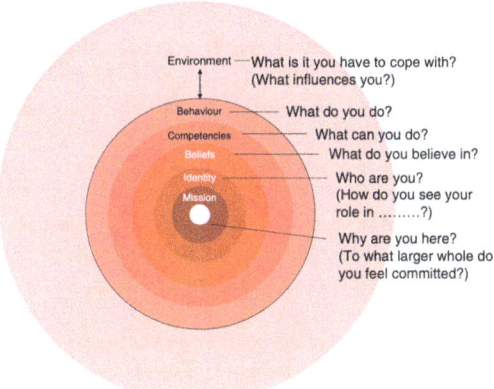

Figure 5.5: The onion model showing six levels of reflection, taken from [120].

If you feel that reading all these blogs and books is too much, there are several options, like watching video summaries of books (https://www.productivitygame.com/books/).

I am not claiming that a book or method I presented will particularly match your needs at any stage in your journey. Nor are any of the other examples or references I am giving you here aimed at providing a silver-bullet solution. Honestly, I do not even manage to follow all the suggestions of any cool or good book.

I believe all the messages from the literature in this category of time management or professional orientations are all trying to show or demonstrate similar aspects with different perspectives and offer complimentary tools.

It is up to each of us to continually try different methods or approaches until we find one that works for us, at least in the particular stage you are. In a few months or years, you may be certainly facing new challenges, keep flowing!

5.7 To lead or to follow?

Not everyone needs to be entrepreneur or leader of a given project. It helps, however, to identify and know the basics of what it means. Being a good follower is as important or even more so than being a good leader. In fact, good followers have better chances of becoming good leaders, because they can learn from their own mistakes, and those whom they have followed. But, there is always a moment where you may want and need to work alone to be creative.

One of my greatest idols is Isaac Asimov (1920–1992). Asimov argued that isolation is required for creativity to pop up. Creative persons are continually working at it, shuffling thoughts and information, even when not doing it consciously.

The presence of others can only inhibit this process, since creation is embarrassing. For every new good idea you have, there are a hundred, ten thousand foolish ones, which you naturally do not care to display. Isaac Asimov Asks, "How Do People Get New Ideas?" A 1959 essay by Isaac Asimov on creativity [8].

He went on to propose that the purpose of cerebration sessions may be to educate the participants in new facts and new combinations, and not necessarily to think up new ideas. As he put it: one person may know A and not B, another may know B and not A, and either knowing A and B, both may get the idea – though not necessarily at once or even soon.

The following question is often asked: *how to persuade creative people* to work together or hold cerebration meetings?

Asimov's answers: First, there must be ease, relaxation and a general sense of permissiveness.

The world in general disapproves of creativity, and to be creative in public is particularly bad. Then, *if a single individual present is unsympathetic to the foolishness that would be bound to go on at such a session, the others would freeze.*

*The **unsympathetic individual** may be a gold mine of information, but the harm he does will more than compensate for that. It seems necessary to me, then, that all people at a session be willing to sound foolish and listen to others sound foolish.*

5.7.1 About leaders and followers

The Economist cited a study [117] highlighting cognitive skills, operational nous and financial knowledge as prerequisites for leaders to succeed. These descriptions show more emphasis on social skills as an ability required by bosses to coordinate and communicate with multiple people.

Thus, a good leader is not necessarily who micromanages to the smallest detail what and how things must be done. In contrast, good leaders "understand the firm's goals and toil together effectively (...) Social skills matter more when bosses *need to persuade* as much as instruct" [62].

The explanation for this shift towards social skills is given as a result of the rise of knowledge workers, e. g., developers, data scientists and other experts that typically work independently – see more in Section 4.8.

In the same line, typically managers or company leaders regularly engage with politicians, address requirements from activists and manage social media attention. The Economist's article ends by listing the abnormal demands for modern CEOs, which also apply to any engineer leader of innovation efforts:
- Be more talented than others in the firm, but don't tell them what to do.
- Crush the competition while exuding empathy.
- Listen charismatically.
- Be "likeably" aggressive, without showing it.

This is a funny mix of functions requiring weird set of skills – like the engineers of the future.

There is another link between the role of empathy and leadership, which I just brush in what follows. Two authors coined the term "Chief Empathy Officer" with the aim to propose CEOs to act with empathy because of the effects brought up by the pandemic, mainly global turmoil on businesses and the workforce. These authors assembled a practical guide to building, using and deploying empathy as "an organizational superpower" [183].

They define *sharing*, *thinking about* and *caring about* as three *core elements of empathy*, matching with emotional empathy, cognitive empathy and empathic concern or compassion – see Section 2.3.5. In that work, cognitive empathy is defined as instrumental when people come from different perspectives, in the context of organizations such as companies.

Besides empathy, authenticity, vulnerability and transparency have been identified as skills for the future leaders: "This is the time for human beings, not human doing." This was said by Leena Nair, Unilever's HR chief [9]. "The half-life of a skill is two or three years. You have to continually relearn, unlearn, reskill yourself." I could not agree more with Leena's statement.

There are ways to characterize leaders: filtered (chosen by the system in the company) or unfiltered (entrepreneurs). Unfiltered leaders are typically unpredictable and induce change because they were not trained in a given system. Each has their place, and arguably unfiltered ones are better when changes or innovation are needed.

There is yet another link between innovation and empathy. A study showed that when people reported their work place leaders to be empathic, they were more likely to report capacity to be innovative [138, 171].

Therefore, it is important for you to identify your "signature strengths" (see page 114) and aligning yourself with the place where you want to work. For this purpose, reflecting on your past experiences helps you find out those strengths (see Section 5.6). What do you keep when do you stop doing things that are not contributing to your WOOP implementation (see page 101)? This is an interesting point of attention where you need to balance your "grit," or your persistence when trying to achieve your goals or dreams.

Another interesting quality that deserves a separate analysis is humility, related to the ability of being humble.[12] After discussing with Dave Blivin, a seasoned venture capitalist, he pointed out that the ability to be successful building a company can be enhanced when the leaders of the venture have the willingness to consider the opinions or help of others. Very often, smart people such as founders of companies – including many engineers I know – think they already know what is needed or assume they must project this image as the leader of a successful company. See page 68 for more on humility.

5.8 Watch out!

I want to highlight a very recent example, and you will have to excuse my extended focus on this particular "hot case" because of three reasons:
1. it is highly relevant to demonstrate the power of teamwork in entrepreneurial activities, particularly when a problem affects the lives of millions of people;
2. it concerns a controversial case that made it (sadly enough) to many news headlines, and she was an inspiration source for many people as the youngest female self-made billionaire; and
3. there are several aspects in the story that can help you avoid common pitfalls in entrepreneurial quests, as well as specific risks related to subcultures or specific behaviors that affect innovators at the global and local scale.

Facts of a top dog story

Figure 5.6: Elizabeth Holmes. Attribution: Tali Mackay at English Wikipedia (https://commons.wikimedia.org/wiki/File:Elizabeth_Holmes_2016.jpeg). This file is licensed under the Creative Commons Attribution-Share Alike 4.0 International license.

---

12 The formal definition of humility is "the quality of having a modest or low view of one's importance: he needs the humility to accept that their way may be better," Apple dictionary, Version 2.3.0 (284). However, I would not suggest you to fixate your thoughts on the synonym "unassertiveness" or other less positive interpretations.

Elizabeth Holmes (1984–) is co-founder and former CEO of the now defunct Theranos health technology company, see Figure 5.6. She left unfinished studies at Stanford's School of Engineering in the USA to pursue an entrepreneurial path. On 4 January 2022, Holmes was found guilty of four counts of fraud, and not guilty of defrauding patients. Holmes did a great job raising funds and demonstrated a great persuasiveness to recruit people to work for her company; however, her dream of testing blood with a few droplets was not fulfilled.

> Please, do not "fake it till you make it."

As recently reported, "some startup founders think it's okay to stretch the truth about their products. That's a dangerous road to go down. (…) The strategy of imagining a future and then creating it – and hoping no one notices the messy gaps in between – is as old as Silicon Valley. A microprocessor, a website, a smartphone: These things all seemed impossible until someone built them" [167].

Theranos stood for revolutionizing healthcare in the analysis of blood, democratizing diagnosis, with benefits for all. Elizabeth joined the mystic list of famous co-founders of companies after dropping out from her university studies. She then filed a patent protecting the idea of a wearable monitor with microneedle to detect biological signals and deliver therapeutical drugs. She got investment in a snowball effect, attributed to contagious apparent confidence of well-seasoned investors.

Many observers were thrilled to see a young and charismatic female founder thrive in a male-dominated field. Her *persuasiveness* has been praised, and it is argued if she was a victim of the surrounding environment. Elizabeth's vision was shared by many, because they thought it was a good idea. Some point fingers to the negative influence of a former boyfriend and chief operating officer (COO), Ramesh Balwani, which led to a criminal scheme defrauding several stakeholders, in particular patients who relied on the tests supposedly performed by Theranos' products.

You can learn many more details from the book *Bad Blood: Secrets and Lies in a Silicon Valley Startup* [59]. I also recommend The "Valley of Hype" behind the rise and fall of Theranos (https://www.youtube.com/watch?v=to2GSibbrv0). The documentary's creators show how an aspiring inventor can define a reality in the future, making a new system or technology, and put the focus on a factor that is very relevant: location. The place where you are can determine the access to capital and lead to specific cultures or behaviors.

"There were lots of arrogant jerks in the semiconductor industry, and part of the culture of the chipmakers was, if someone was producing amazing technology, you were kind of going to let them get away with murder." – Margaret O'Mara, Professor of History, University of Washington, The "Valley of Hype" (https://www.youtube.com/watch?v=to2GSibbrv0), min 15:10.

"Stanford's biggest most important product isn't tech, it's *people* (…) you have access to incredible resources, professors, business people coming to campus, modern Stanford has be-

> come this town square for billionaires (…) where you rub shoulders with incredibly successful people and they are telling you 'this is how you can be like me'." min 19:24.
> I think she is afflicted with what the police call noble cause corruption, meaning (…) the end justifies the means.
>
> Alex Gibney, director of The "Valley of Hype"
> (https://www.youtube.com/watch?v=to2GSibbrv0), min 32.26

Shockingly now, at least for outsiders like me, the early Board of Theranos had almost no experience in diagnostics or any activity central to the main purpose of the company, such as biochemistry or molecular biology – see Section 5.10. I wonder, what can former generals and politicians add to a start-up on diagnostics trying to scale down or miniaturize the sampling of blood for instant analysis?

I see this behavior related to what Perez-Breva alludes to about how "so-called innovation recipes" over-train innovators to focus on the user, when an innovation is in the making [148]. In many pitches I have seen founders or students talking about a product that does not exist as if it was already a reality.

In hindsight, companies like Genalyte, Athelas and Truvian are doing now parts of what Theranos was supposed to do. This means that Holmes had the right concept and the right story, but the technology was not ready at the right time. I believe that being transparent about crucial processes, *like some empathy communicating to stakeholders*, should be a top priority for all entrepreneurs. The bottom line is that if Theranos would have been more honest about what it could and what it could not do, this story would not have ended so bad.

A lack of transparency and going too fast can kill the best of ideas. The intrinsic desire to bring a solution to society as fast as possible is accelerated by the "hype machine," namely media attention and a public hunger for praising champion celebrities. Typically, technological development progress is moderated by the slow pace of peer-reviewed studies.

But sometimes a company may opt not to publish all data to limit competition. "Stealth research," as Prof. John Ioannidis from Stanford University defines the secret way of working, stands opposite to peer-reviewed research. Peer-reviewing increases trust in results, because they are scrutinized by other experts and must sustain criticism. In contrast, media and investors are in short supply of time and often lack the *knowledge* to check the claims of the companies they are investing in.

> "When you start a new science project, you don't know what the unknowns are and so that naïveté of not knowing what roadblocks you're going to hit, what you're going to have to solve along the way, I think that has a long way to go well with being an innovator, the naïveté joke is a really important piece of innovation."
>
> Cary Gunn – Founder of Genalyte, a blood diagnostics company in The "Valley of Hype"
> (https://www.youtube.com/watch?v=to2GSibbrv0)

The documentary ends with a quote from Holmes that resonates with what many other founders of companies and professionals that are convinced of what they are trying to do:

> "If I were fired, or I had to start this company over a lot of times in order to figure out how to get this right, I would, right?"
>
> Elizabeth Holmes, quoted in The "Valley of Hype"
> (https://www.youtube.com/watch?v=to2GSibbrv0)

I invite you to think about which answer you would give in a similar circumstance.

Top dog KPE analysis

I find it interesting to contrast the KPE analysis I provided for the example from the nineteenth century in Section 2.5 with this latest case. Contrary to that case more than 100 years ago, the amount of detail for Theranos' ordeal is just overwhelming. Thus let us stick to the summary I provided about Holmes' fall from grace for this analysis I give in the context of KPE.

Knowledge How much *knowledge* did Holmes have at each step in her journey? I do not mean technical knowledge necessarily, but also knowledge on how to treat patients or drug development. How much critical thinking and self-criticism on ethics did she have? How aware was Theranos of the actual levels of honesty and integrity in the team?
- Most engineers are aware that (over-)optimistic ideas not always work in practice. As an entrepreneur she may have pushed for invention or fraud (aware or not).
- What makes it a group mistake in my view is that all statements made by Holmes and Theranos should and could have been corroborated by the investors and several partners. Instead of checking the facts or gaining more knowledge, lots of money flowed to Theranos for more than a decade and little questions were posed.

Persuasiveness The fact that Holmes had good persuasive skills should not be questioned considering how much money she raised. From my basic understanding, it seems Theranos, Holmes included, promised more than could be proved at the specific moment when they were raising funds. We can only speculate how easy it was to persuade her to do the things she now denies knowing about. For example, there are allegations that her former partner and COO of Theranos, Balwani, was very controlling and manipulative.

Empathy However, how much empathy did she have for the stakeholders? Was she consciously taking part in a culture known to have certain flaws to make sure her dream or ambition of making a difference in the world would be a reality? It is

possible that with the right coaching or a balanced board of advisers, she could have changed the course of actions.
- The worst of it all is that so many people's expectations were unmet, and the disappointment caused to patients and the damage to the public's opinion about entrepreneurs are immense and long-lasting.

The ultimate question is, how can we distinguish between fraud and visionary statements? If a good liar and excellent storyteller can get funding to develop medical devices and affect people's lives, who can stop it? I propose that by stressing KPE to the max, we may contribute to educating all stakeholders and "take nobody's word for it" (see the quote at the beginning of Chapter 3).

Then comes the thorny issue of how we measure and treat *success* – see Section 5.11 for more on this. If Theranos had delivered on all that was promised, nobody would have been so focused on this story. It may be argued that charisma or persuasiveness is what allowed Elizabeth to go so far, and that is not bad in itself. The main deficiency I see is a lack of competence of the whole innovation team (executives, board of directors, etc.). Unfortunately I have no means to reach Elizabeth to learn from her personal journey, as you can see in Section 6.2.

Theranos had a good idea and lots of money, but a more balanced team with a bit more empathy could have avoided this unfortunate sequence of events.

5.9 About dog-eat-dog

Here I want to shortly touch upon two important elements. As you probably noticed, Sections 2.5 and 5.8 include KPE analyses and references to the terms *underdog* and *top dog*. This allows me to draw your attention to:
1. the need to extend empathy beyond humans (in this case, animals); and
2. the need to rethink the whole idea of "survival of the fittest."

Frans de Waal, who has studied apes for 50 years, has exposed why "survival of the fittest" is wrong [10].

> "People sometimes describe nature as a dog-eat-dog world. Some of the biologists depict nature as a battlefield basically where selfish tendencies tend to prevail. And from morality, the evolution of morality there's very little room. What they mean is that all they see is competition. I win, you lose, winning is better than losing and so on. That's totally wrong. I fought against that sort of characterization of animal society all my life, because just like human society is built on a lot of friendship and cooperation at the same time. We'd like to deny that connection that exists between us and animals. Certain tendencies, such as a sense of fairness, *empathy*, caring for others, helping others, following rules, punishing individuals who don't follow the rules, all of these tendencies can be observed in other primates. And they're saying these are the ingredients that we use to build a moral society."

That being said, we have now set the ground to discuss the (obvious) need to consider work in teams.

5.10 About teams

A venture capital expert once told me that for a start-up company to succeed, there are three elements to pay careful attention to:

Idea: This is kind of obvious, because almost everyone in the planet thinks their ideas are good, and many truly believe all their ideas are "the best."

Money: That is a "no-brainer," because you need it to pay salaries of the people in the team and outsourced services, purchase equipment and gain access to knowledge by buying books, for market analysis, etc., *ad infinitum*.

Team: This is then the real deal-breaker, because the composition of a team can really tilt the balance of any venture.

Having one of these three elements may not get you too far, but with two of them you have great chances to make good progress. But as we have discussed in several parts of this book, humans and their actions – as well as ideas – are what move the progress of science and technology, for the betterment of society.

I hope you can agree with me that to solve most urgent engineering problems faced today by our society, good collaborations are required. With some exceptions, perhaps a mathematical problem or a computer simulation, I cannot think of solutions that are not based on tools and knowledge coming from different disciplines. The literature in this regard is abundant, but I thought it relevant to touch upon a few points that can connect to the *CBL* approach introduced in Section 4.4.1 and other activities needed to innovate.

If you are interested in solving big problems, such as poverty or the transition from fossil fuels to renewable energy, having only technical knowledge will not get you too far. If you do not collaborate with economists, experts in humanities and the actual population which will benefit from your solution, there is a very slim chance you will make it a reality.

I have not mentioned politicians until now, because in most countries, politicians come from different backgrounds. However, it is often said that if there would be more engineers or scientists involved in politics, or at least politicians that would listen to them, we would be in a much better world.[13]

But as a teacher and scientist, I cannot influence much the current state of affairs in politics. Therefore, let us go back to what I can teach you better. In some of my courses I invite my students to answer a few questions from the team roles test (https:

---

[13] You may want to check the note with Devaraj's opinion on page 23.

//www.123test.com/team-roles-test/), based on The Nine Belbin Team Roles (https://www.belbin.com/about/belbin-team-roles). Research shows that the most successful teams are those where different behaviors are present: Resource Investigator, Teamworker and Coordinator (the Social roles); Plant, Monitor Evaluator and Specialist (the Thinking roles), and Shaper, Implementer and Completer Finisher (the Action or Task roles).

The interesting experience I have with these roles is that if you take the test now and repeat it in a few hours or days, you might get a different result. At the beginning I doubted this lack of reproducibility, but soon I concluded that these tests and roles should be used as a guide of behaviors or attitudes that are needed at different stages in a project, and you do not need to fit in any particular role.

What I suggest my students when arranging a team to work on a project is to try to find out which of these roles they feel more identified with, and try to see who can fill up the missing ones.

In the real-life arena, diverse company founding teams are likely to achieve high performance in competitive commercialization environments, while technically focused founding teams may achieve success when the enterprise pursues an innovation strategy [77]. Also, in my scientific research, I have experienced how valuable it is to have people with different knowledge backgrounds, as well as cultural experiences and ages. Look at this varied team composition as directly contributing to increasing the applicability (see equation (2.1)) of any project you are working on. More on this is discussed in Section 4.3.

The book Scale [180] has examples of how analogies and tools from different fields can be put to good use. One of them is the program Information Society as a Complex System (ISCOM), focused on problems ranging from innovation to information transfer in both ancient and modern societies.

One of the questions studied was whether cities and companies manifest scaling and devising a quantitative principled theory of their structure and dynamics. I found the following part interesting, where a retrospective look is provided on how teams can be formed.

> (...) if I look back at the list of attendees at one of our early workshops, very few of them eventually became ongoing members of the collaboration. This is not unusual at the beginning of a program such as this that proposes to broach new questions that transcend disciplinary boundaries. At the outset all kinds of people with diverse backgrounds who are well versed in expertise that might be pertinent to the program are invited to participate in the hope that synergies will happen, sparks will fly, and a real sense of purpose and excitement about prospects for something new will be generated.
>
> Many find, however, that even if they are fascinated by the intellectual challenges and potential outcomes of the proposed project, it simply isn't compelling enough to sacrifice the time to get fully involved and reset the priorities of their own research agendas. Others discover that they really aren't that interested after all, or that it's unlikely that anything of substance will come out of the effort.

> Eventually, however, by word of mouth, by serendipitous connections and informal discussions, and by osmosis and diffusion, an evolving group of researchers emerges whose members are to varying degrees willing to commit to a longer-term involvement with the challenge and who will actually do the substantive work over the ensuing years.
>
> Scale [180], page 241

That analysis contains the silver-lining of all that we have been discussing in this book and focus more on in the last section of this chapter.

5.11 Success or failure, what is the question?

The discussion about leaders and followers from Section 5.7.1 connects to how you want to define success. A lot is discussed in books and the social media about who is successful and how much failure was part of the success. Not having what you want when you want it does not need to be seen as failure. Not achieving 100 % of the goals set in a given period does not need to be considered a failure.

> "Saying you feel disappointed by something you want but maybe can't have makes you honest... And admitting your vulnerabilities makes you courageous."
>
> Otis Milburn, main character in Sex Education, Season 3, episode 8, Min 39:27, British comedy-drama streaming television series created by Laurie Nunn for Netflix

My view on this is that you should definitely look around, get inspired and learn from others, but it is more relevant when you "compete" with (and not against) yourself. You can definitely hold yourself accountable if you decide to skip your homework or responsibilities in a work project, in exchange for "other" activities. It is almost futile to try to compare your journey with someone who lived in very different situations than you, e. g., years before, in a different country, working on different problems or facing other biases than yours.

That is why I find it a bit uncomfortable when people want to imitate or try to claim they are the next *XXXX YYYY*.[14] Instead, I propose you look back at your goals from a few months ago and go through the different models or acronyms you feel comfortable with, for example WINE, WOOPS or KPE tips from Section 5.2. Doing this regularly – see also the *Log progress* step on page 89 – will give you a good indication of your success and fulfillment.

---

[14] Fill in the famous inventor or scientist's name of your choice.

5.11.1 The value of time

Many members of the public reacted very surprised when it was announced how effective the vaccines prepared to fight COVID-19 were developed in apparently such a short time. This is analogous to the tendency of seeing only the success stories or the moment to shine. However, very often those success stories are the result of long years of work. Most developments in science have not been as revolutionary as is written or remembered. Instead, it is an evolutionary progression based on existing technologies and knowledge accumulated until that point. To make it clearer, a saying in the technology industry world states that an *"overnight success takes years to create."*

The fact that *time* is typically required to succeed is no coincidence. During my conversations with professionals of all disciplines, entrepreneurs and experts in valorization, I noticed some common characteristics that can help you grasp how valuable time is. Ultimately, it is all about what you can do to make "failure" work in your benefit. In other words, *with the appropriate storytelling, failing looks better than it actually was.*

Sometimes, a flaw in one historical moment or place can become the required quality. Here I give you an example that connects to my favorite sport, swimming. Michael Phelps' biography [17] indicates that Phelps has a very long torso and "short" legs. His looks probably generated jokes during teenage years: he is even known as the "Flying Fish." In fact, the disproportionately large chest of good swimmers enables them to swim with less drag or water resistance from shorter legs. There is a longer list of the reasons why Phelps has succeeded, including the power of his mind, but we better get out of the water and continue.

In STEM, we are typically associated with being "geeks." The Geek Syndrome [18] discusses how autism and Asperger's syndrome and its surge among the children of Silicon Valley could be correlated with math-and-tech genes. I do not see value in entering a discussion that is full of suggestions or leads about whether environmental factors such as industrial pollutants in air and water and certain foods are responsible. My intention is to encourage you to accept how you are (even if you think you have many flaws) and try to grow on qualities or strengthen specific skills that will help you stand up among your peers and eventual competitors.

To close this part, you could try to compare your flaws against *signature strengths*, and here I leave two places where you can find more about it: Verywell Mind [19] and VIA Character [20].

Now, I want to provide a selection of cases where time "plays" an important role.
1. Most "hardware" inventions take more than approximately 5 years, typically 10–20 years, to generate revenues or profit.
    - Interestingly, the delay in adoption or proofed success is the result of different factors, but irrespective of the sector, activity or country, it leads to the

same timeframe, for example in chemical engineering or for biomedical products.
- For some innovations, e. g., a new drug, a new drug delivery method or a cool invention within a big transnational with a heavy bureaucratic and self-censorship apparatus that avoids disrupting successful products, it can be the regulatory aspects. Imagine a new reactor that could replace a relatively new or very profitable reactor or process; managers may not want to change a good technology unless the advantages are too good or some regulatory aspect or competition element persuades them to do so.

The discovery that became the basis of the electric motor propelled Faraday's name across Europe and later the world. The pace of technological change was much faster in the nineteenth century than ever before, but it still took 60 years until Faraday's prototype – a wire circling around a stationary magnet – was used to run electric trains in Germany, Britain and the USA [104].

---

2. The most famous universities or institutes with most impressive indicators of "success" undoubtedly have plenty of smart and highly motivated people.
   - In fact, what makes their apparent success is not the fact that they are smarter than anyone in a smaller university or less developed country. The best groups or innovation systems work by typically trying many – seemingly crazy – approaches, failing fast, learning from their mistakes and moving on to the next opportunity (saving lots of time if you are limited in resources such as funding). See also the quote on Stanford on page 107.
   - The morale of this story is that if you do not have that much access to resources, time is your best ally. You could even "cheat" by collaborating and distributing the workload. This hinges on the assumption that you keep good memory (notes, publications, etc.) of what you have done and try differently next time. Sometimes I call "recycling" of ideas when I try a given proposal or solution against different funding agencies; see the example in the next note box.

One of my most painful and enlightening professional experiences was applying for research funding right after finishing my PhD. The Dutch Scientific Council has a very competitive personal grant scheme, and one of them is meant to help scientists in the early years after becoming doctors to follow their own research idea, the VENI (https://www.nwo.nl/en/calls/nwo-talent-programme). In the years I applied, it had a success rate of around 15 % or less, and it involved several steps spanning almost one year: submission, peer review, rebuttal letters and an in-person presentation in front of a jury as the last step. I went twice to the interview, and never got it.
The first time I did not get it was tough, but I knew I had another chance and the criticism and suggestions for improvement I got were reasonable and easy to implement in my new attempt a year later. For example, my track record or CV got stronger and the basic idea gained more evidence of applicability.

The second time, the comments were completely different, and I got very frustrated, because I took it as a personal attack or at least I did not identify myself or my proposal with what the jury report was telling me.

With time, I learned how difficult it is for a jury to order priorities when assessing very good ideas from ambitious scientists. There were some jury members from the year before, but I had no clue if the reviewers were the same. I also realized that there is a luck component that you can never really control in these sort of competitions.

For example, if someone in the jury does not like one particular aspect, or one of the reviewers gives a negative comment, there is no time or resources to double check it all – similar to what investors and board members face when working on promising ideas; see Theranos' story in Section 5.8.

The morale of the story: "do not take personal competitions personal." That is actually a joke I like telling, but it basically says you need to focus on the idea or product and not get affected by subjective assessments from a particular jury, funding agency or investor.

Let your ideas grow and sharpen your storytelling skills. When competing, ask trusted colleagues to read your drafts. Not only does it help improve your idea, it also lets your peers assess the quality of your ideas and they will consider you for future projects.

There is a happy end to this story: the basis of my idea eventually made it to a successful project that we are now running.

---

3. Most inventors or innovative thinkers fit the classical stereotype of clumsy and chaotic, or, more elegantly put, eclectic.
   - There is much true to it, but there is an almost hidden element. When the moment to make the links between disparate sources of information comes forward, they convert the information at hand, mix it with old knowledge and align all that chaos towards a solution: new knowledge, bang!
   - This is what I call "radar vision," and it is a very effective quality that can be trained – the opposite of tunnel vision.

   You can compare it with design thinking and other ideation processes, where you first think of solutions during a divergent phase, assessing different avenues or options, followed by a convergent phase that leads to the moment of choice, or choosing a solution.

To wrap up this section, the ongoing constant changes in our world can lead to anxiety or frustration, discussed briefly in Section 4.8 and the note on page 86. The negative side of this mental state can be mitigated partially by realizing that the passage of time can play in your favor. Nobody wants to be recognized for their achievements after it is too late,[15] but I would suggest to take a different look at it.

When we are starting a project or have a burning drive to solve problems, there is an impulse that pushes to accelerate things. Sometimes things go very well, and rarely we stop to think about how lucky we are. In fact, most projects I have worked on never

---

[15] Think about all those posthumous recognitions many famous inventors never enjoyed.

go exactly as expected or planned. Know that if one idea or solution you have does not "catch on" instantly, you may have to wait a bit, instead of trashing it right away.

But do not stand still while you wait! Make sure you actively engage with as many stakeholders you can reach, not only in direct connection to the topic of your invention or idea, but on a broader spectrum. Speak up, write a blog, engage with younger and older audiences.... Sooner than you may know, your idea will fit the right audience, and you may catch a wave, because Hot Streaks in Your Career Don't Happen by Accident. First explore. Then exploit [11].

I believe the ultimate challenge for you as a professional is to ensure you find your balance or The Science of "Flow States" [12]. Flow occurs when the challenge of a given task and your skills at the task are both high.[16] Also, loop around your ongoing project or projects as implied in Figures 2.12, 3.2 and 5.1.

The list of tips I could give you may be endless, but I think we better stop here. I am sure there are plenty of other tricks and ways of working you can explore on your own or within a team, and the internet nowadays is full with useful (and sometimes distracting) material – check the rabbit hole in Figure 1 in the Preface.

Just do not settle on the beaten track or stick to a tunnel vision. Find contrasting opinions and try to *empathize* with the different people and the views you encounter. Match the different messages or suggestions from them with your goal, whatever it may be at that point. Persuade and allow to be persuaded if needed. Try to work, even as intern, in early-stage companies, or if there is none around, join a group from a research institution. Summing it all up, following the KPE framework will enable you to understand what it means to be an empathic entrepreneurial engineer.

---

[16] If you like (video)games, that is a good example of flow experience [67].

6 IEEE: Interviewing empathic entrepreneurial engineers

Reading time ~ 30 min

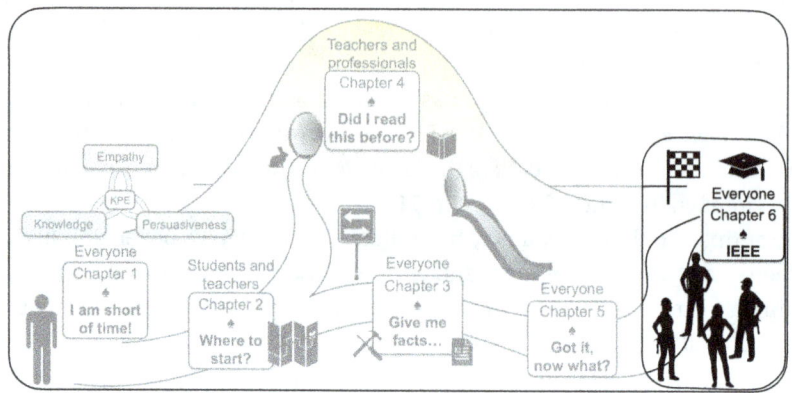

Bit: Snapshots of other people's journeys

> Mathematical reality is an infinite jungle full of enchanting mysteries, but the jungle does not give up its secrets easily. Be prepared to struggle, both intellectually and creatively. The thruth is, I don't know of any human activity as demanding of one's imagination, intuition, and ingenuity.
>
> Paul Lockhart in Measurement, p. 2, 2012, Harvard University Press [125]

This chapter contains what I call mini-interviews I conducted online to some of my collaborators. I pitched them the idea behind this book, and they have shared parts of their professional trajectory through the lens of KPE.

In the early months of the COVID-19 pandemic, around February 2020, I had to "work from home," as many knowledge professionals. My first thought was:
"oh shoot, what about all those BSc, MSc and PhD students?"

For seasoned professionals, or tenured professors like myself, having a gap year or a few months disruption in their CV may have certainly an impact.[1] But if your study program is typically designed to last for three years, sitting at home for even a third of the time is an even bigger change.

---

[1] In the scientific sector, the impact for women, and parents of small children, was the worst. See this article about the career cost of COVID-19 to female researchers, and how science should respond [41].

https://doi.org/10.1515/9783110746822-006

6.1 Empathy-driven need to write

Despite advanced technologies for remote teaching and working overtime, the quality of the education has been affected, crucially at an age where students should be building relationships with other fellow students and teachers. It is now when students should not only be sharpening their technical and professional competences, but also gaining and nurturing the "durable" skills.

In contrast, all that we all had for painful months was a cold video-conference application to interact with the world, in the best cases.

Then my "empathic" brain started thinking on ways I could help all these "engineers in the making," waiting in their pyjamas, fighting the temptation to go to online streaming or gaming services.[2]

The result from my resistance to watch too many video streaming services is this book I am so happy to have written for you – particularly if you got this far! The fact that I was not 100 % successful in not watching some series and documentaries becomes evident along the book, peppered with some popular cultural and historical references, particularly in Chapter 5.

6.1.1 Distilling ideas

The last two years have made a great impact on my personal life and professional career. With this book I channeled my energy and frustration of being locked at home, and tried to focus on a message that I believe can help the future engineers, and well-established professionals, too.

This book is the result of a deep self-retrospection and meditation about my role as engineer and my best wish to help other people who are willing to improve the world. In the process I ended up with a proposal to update a definition of what an engineer is today (see Section 2.3.1) and the concept that I developed into the book you are reading now: KPE.

An engineering student in the classroom right now – online or in person – is learning about technologies or knowledge that might not be sufficient to tackle the ever fast changing challenges of our time. It is simply too much to learn in a three- or four-year BSc study program. It certainly does not improve much by gaining extra degrees (MSc, PDEng, PhD, ...). We may spend hours arguing about the minimum requirements to call someone "engineer" and what type of skills this person must have to be allowed to perform as such.

You may agree to a certain extent that empathy is a quality some people possess through personality or through upbringing. To my (happy) surprise, I found I was not

---

[2] I must confess I struggle with the temptation to watch science fiction and history dramas.

the first one in the engineering disciplines thinking about it, and it is being taught in a handful of STEM study programs, with varying levels of awareness. The actual interpretation and its importance in different professions are matters still discussed by academics, and lately in the professional literature, as I briefly present in Sections 2.3.5 and 5.3.

This quality, in my opinion, can make a big change in the personal development of engineers, the economy and the environment at large, as discussed in Section 4.3. This quality may be considered by some as "soft," but I came to know through communicating with colleagues more experienced than myself that we should replace the label with "durable" to help building resilience for the crises to come (see Section 2.3.4).

In fact, most of what we learn as social skills may outlive many technological developments we enjoy today. Depending on your age, you may not even know how computer programs were run with perforated cards a few decades ago. My first programming experience was with a magnetic tape recorder around 1991. Thus, if we are to focus on the minimum requirements, "durable skills" should be on the top of any list.

When I reached this conclusion, I wondered for a while thinking about:
1. How can we teach it?
2. Can it be done via online tools?
3. Once we find how to teach it, can we also do it outside the classical educational settings, such as in companies and other types of organizations?

Moreover, if we are to educate remote communities and bridge the yawning development gap between different parts of the world, we need to enhance the empathy in all people – see Chapter 4, where I share my proposed approach to achieve this.

Already before social distancing was enacted in 2020, there was a tendency among engineers to no longer work necessarily at the same time in large groups, e. g., bureaus, firms, governmental labs, etc. – like many knowledge-economy professionals. Even before this worldwide transition to freelancers or the "gig economy," it was common to have mentors, formal or informal, or more experienced colleagues who could help the newcomer in understanding how their work fitted in the "larger picture," e. g., customer needs, interdepartmental dynamics, and so on.

Being "away" from my direct colleagues, students and family members living abroad gave me the chance to look back at my own trajectory, and I realized that empathy is perhaps the most important and durable skill I gained.

The "soft skill" label does no merit because I think it is very "hard" to acquire or master. The good news is that we do not require expensive equipment or talk to busy or very important people.

If you read this book – or at least some parts – you will find several examples and real-life cases that can teach you the basics.

Next, I started to think about how I could help those younger students or recently graduated engineers with reduced social interactions. The ability to connect to our fellow human beings is something that can give the right motivation and align our efforts towards a real solution to any problem we can think of.

Doing it, empathizing, can help some experts or novice professionals to descend from their ivory towers and use their skills or knowledge to a better aim. It can open our eyes and help you realize that perhaps we were making the wrong assumptions when working on a technological problem and forgot what impact any solution may have, not only on people, but on our environment as well.

Therefore, having identified a "solution to an urgent problem" I summon you, reader of this book, and all engineers to join me in the challenge I took upon myself: to connect better to the newer generations of engineers, teach them and promote empathy among all engineers.

My first step in this direction was to reach out to my colleagues and students to discuss the KPE idea, and we ended up having dialogues or mini-interviews that you can read in the next section (Section 6.2).

6.2 KPE seen by other innovators

This section contains what I labeled mini-interviews given by innovators in different areas of great societal relevance. I mentioned before that sometimes it is hard to find analogies or get actionable insight from the stories told about innovators from the past, because so much has changed (see Section 5.11). Therefore, I thought that talking to entrepreneurial engineers risking their time and energy *right now* could make you think of yourself as a potential next innovator.

Wherever possible, we provide a simple *IF* analysis to show you how it can be applied beyond the examples from the previous chapters.

6.2.1 Artificial intelligence and chemistry

Daniela Blanco
Co-founder & CEO Sunthetics
Daniela obtained her PhD in Chemical Engineering from New York University (2020) and is the co-founder and CEO of Sunthetics. At Sunthetics, she develops machine learning platforms that accelerate chemical innovation for more sustainable manufacturing in the chemical industry. Daniela has multiple patents and awards for her technologies, including Forbes 30 Under 30 in Energy (2021), Top Innovator Under 35 by MIT Technology Review (2020), Top Female Founder by Inc Magazine (2020), Best Global Student Entrepreneur by EO (2019) and Brightest AI-CI Mind by the MIT Center for Collective Intelligence (2019). She has also been featured by Disney+ [141].

We accelerate sustainable innovation in the chemical industry

The production of chemicals is indispensable in the products we use on a daily basis, from electronics to textiles to food. The chemical industry, however, has become the third largest contributor of greenhouse gas emissions, with more than half of its resources ending up in waste streams. Sunthetics' mission is to make the chemical industry more sustainable, one reaction at a time.

The only way to create lasting sustainable impact in the industry is to fast-track current innovation around processes that have *lower energy consumption*, better *resource usage* and *safer* or bio-based starting materials and *integrate renewable energy sources*. We offer a machine learning tool that is capable of leveraging small datasets to generate big insights. Sunthetics' platform enables process chemists to innovate and pinpoint viable processes using only a few experiments, accelerating the path to market for innovation by 75 %.

 Was "empathy" ever taught or discussed as a relevant skill you needed, for either work or your personal life?

Yes, during my personal upbringing (family, friends, other nonprofessional activities) and after university during my professional career. I believe some study programs are lacking projects or courses that encourage students to creatively think of solutions that solve real-life problems. The more engineers and scientists are connected with the real problems of our society, the more we realize that our *knowledge* can make a difference. Our knowledge is not intended to remain printed on a notebook or an exam page, it's supposed to be used with ingenuity to improve our lives. I think this becomes more clear when we approach complex real-life problems that are solved through *interdisciplinary collaboration*, making us realize that we don't need to hold all the answers ourselves and that creativity is key to find new solutions to existing problems.

I also think it is extremely important to be exposed to these concepts *as early as possible*. It is never too late to learn that as engineers or scientists we have the power to drive technologies that address issues that matter to us. I hope engineers will start recognizing sooner in their careers that the impact and success of our work will depend on how it solves specific needs, how it affects our world and how useful our technologies can be beyond our work space.

It was only during my PhD that I learned I could be a leader with my ideas, I could make them happen. Anyone can come up with a new idea and there is an incredible number of resources available to learn what is needed to implement them. Before the PhD, I wanted to work on technologies that could impact our industry, but I had no idea of "how" they will come to have that impact. I thought my job only went as far as to develop the technology. Then, I realized, *my job goes as far as I want it to go*, and I chose to take it all the way to commercialize it and ensure it has the impact I dream of.

How do you see KPE in your journey?

I have constantly leveraged K to gain credibility and performance, P to gain support and E to adapt and ensure our product is solving a true need in the market. Although all three have been key in our journey, I believe empathy has been instrumental for our success. We started our company with a clear idea of how we wanted to inflict change in the chemical industry. However, our customer-discovery activities taught us that the customer had different needs, pains and challenges than those we were imagining. We had major pivots in our technology and business model because we relied on our empathy towards the customers to understand their needs and what the right solution would be.

Without empathy and understanding of the customer, the business has no future. I think empathy has also been key while growing our team. We need to understand our team culture and the candidates that come along the way to join the company. Empathy is a key asset to help us respect our company values and culture as we grow our company.

Miguel A. Modestino
Professor NYU, Co-founder Sunthetics
  Miguel A. Modestino is an Assistant Professor in the Department of Chemical and Biomolecular Engineering of New York University (NYU). Miguel obtained his BS in Chemical Engineering (2007), his MS in Chemical Engineering Practice (2008) from the Massachusetts Institute of Technology and his PhD in Chemical Engineering from the University of California, Berkeley (2013). From 2013 to 2016, he was a postdoctoral researcher at the École Polytechnique Fédérale de Lausanne (EPFL) in Switzerland, where he served as project manager for the Solar Hydrogen Integrated Nano-electrolysis (SHINE) project. He is a winner of the Global Change Award from the H&M Foundation (2016), the MIT Technology Review Innovators Under 35 Award in Latin America (2017) and Globally (2020), the ACS Petroleum Research Fund Doctoral New Investigator Award (2018), the NSF CAREER Award (2019), the Inaugural NYU Tandon Junior Faculty Research Award (2020) and the NYU Goddard Junior Faculty Fellowship Award (2020). His research group at NYU focuses on the development of electrochemical technologies for the incorporation of renewable energy into chemical manufacturing. He is also co-founder of *Sunthetics Inc.*, a startup developing electrochemical reactors and machine learning solutions to accelerate the development of sustainable chemical manufacturing processes.

Was "empathy" ever taught or discussed as a relevant skill you needed, for either work or your personal life?

  Empathy was never taught in my formal engineering education, and not even directly in my entrepreneurship training at Berkeley Haas School of Business or at NYU. Our formal education is built based on hard metrics that are measurable and usually lead to improved performance of a technology or a lower cost, rather than empathy with our stakeholders. Even in our own entrepreneurial experience with Sunthetics, we learned early on that sustainability was a "soft" added value to key stakeholders (investors, large industrial partners), but unless there was a measurable and substantial economic benefit it was not to be a defining factor for our technology. While it was clear that customers do value sustainability, there was a lack of empathy for that customer perception from investors or corporations, which prevents start-ups to grow in the cleantech space. Fortunately, we are starting to see a (slow) change in the value of sustainability, which is driving corporations and investors to be more empathic, which ultimately will increase the value of empathy in the entrepreneurial journey.

How do you see KPE in your journey?

  The path going from discipline – valorization – legal is common. The KPE aspects in our cases are embedded into each of the steps. Even at the discipline level, there is a lot of P and E required to secure research funding, train researchers and get them to focus on a valuable outcome. I think between opportunity identification and real-world entry, P and E are at the same level. Working on the sustainability space I used to focus on arguments to persuade stakeholders (e. g., industry, investors) on environmental benefits. While it was clear that they had an interest on those aspects, it became evident that I had to empathize with their economic constraints to be effective at persuading.

IF **example**

Sunthetics enables the development of new chemicals, products and formulations in 20 % of the time typically required. Our technology reduces waste and emissions in the chemical industry. This directly targets climate change and irresponsible manufacturing problems. Sunthetics' technology is designed to democratize the use of machine learning in the chemical industry. As an easy-to-use, reaction-agnostic platform, Sunthetics improves the productivity of engineers and scientists by accelerating optimization and improving their results.

As a case study, we worked with a chemical company in the United States to optimize a process for the electricity-driven production of a chemical used as a key intermediate in pharmaceuticals and fragrances. The goal was to maximize reaction yield. Following the traditional empirical experimental approach for process optimization, the company ran 39 experiments guided by the intuition and expertise of their engineers. The highest yield found was 86 %.

However, when using our machine learning platform, the company reached a 93 % yield with only nine experiments. In this case, their work was guided by the machine learning algorithms. The pilot delivered an increase of 7 % in reaction yield with 75 % less experiments. Focusing on the yield achieved, the experiments run and

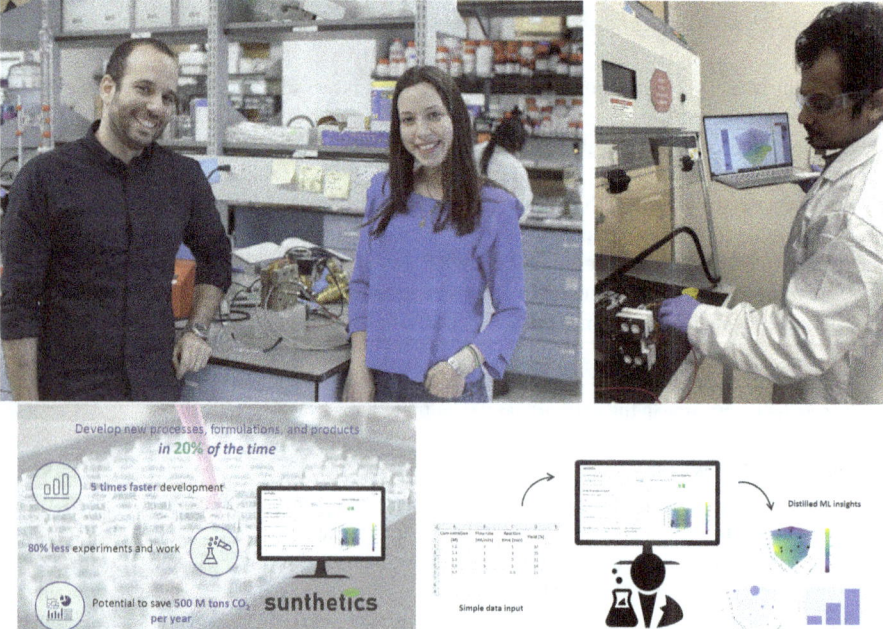

Figure 6.1: Daniela and Miguel stand in their lab; to the right, their AI software in use. Below, you can see the main features of Sunthetics and the logical steps using their platform. Pictures, courtesy of Daniela Blanco and The NYU Tandon School of Engineering.

the time spent in process optimization, we can calculate an *IF* factor of 16.4 (see Figure 6.1). As expected, the *IF* calculation shown on Table 6.1 is larger than one, which is advantageous for our clients.

Table 6.1: Comparison between conventional optimization (Before) and Sunthetics-guided optimization (After) of the electrochemical process.

Case	Factor	Before	After	d	IF
Sunthetics	Yield [%]	86	93	−1	1.08
	Experiments [#]	39	9	1	4.33
	Time [days]	168	48	1	3.5
				IF_{total} =	16.40

6.2.2 A is not always for Apple

Connie Nshemereirwe
Co-founder & CEO Actualise Africa
Dr Connie Nshemereirwe is an independent science and policy facilitator and acts at the science–policy interface as a science writer, trainer and speaker. A civil engineer turned educator, her primary interests lie in creating solutions for the transformation of the education system in her home country of Uganda, especially targeting basic skills acquisition and equipping teachers with the ability to better adapt to changing times. She is also passionate about equipping scientists with the skills to communicate their science better with the public. She is the sole proprietor of Actualise Africa, which she established in 2016.

Closing knowledge gaps

One of the issues that *Actualise Africa* is concerned with is finding ways to make up for the gaps in the learning of children within the schooling system. This journey started out by focusing on equipping teachers in the lower grades of schooling better, but I soon encountered a deficiency in my knowledge of the educational needs and experiences of learners and educators in the lower grades. Thus I have evolved to creating solutions to address the gaps left behind by the schooling system beyond high school. This is an educational level that I have much more experience and knowledge about due to the fact that I taught at university level for 15 years, and my primary research is on the transition from high school to university.

This is in the early stages of development, but what I suspect so far is that most of the people for whom I am developing a solution:
1. are unaware that they have gaps in their skills, especially in their literacy, numeracy and critical thinking skills;

2. currently employ unconscious or conscious workarounds to overcome these deficiencies, for example, by using templates, asking colleagues for help or avoiding certain tasks altogether;
3. having completed a given level of schooling, e. g., high school or university, and having attained a given (acceptable) grade, believe themselves no longer in need of formal instruction on some of the things that they supposedly learned in school, e. g., writing or applying numerical concepts correctly.

One of the parameters of success is whether people, after taking a short pretest, can recognize that their skills levels are lower than they thought they were, and then this would lead to them taking on one of the modules by which they can improve these skills, and then convincing themselves that this skill is now to their satisfaction after they take a self-posttest.

 Was "empathy" ever taught or discussed as a relevant skill you needed, for either work or your personal life?

> I have a feeling this was not even a thing back in '91. But as a result of various encounters with students and fellow educators over the years, however, I have come to understand that my reading of a problem or a reality is often extremely narrow. I was once trying to help a highschooler to perform better in his physics lessons when over time I realized that the problem was not his ability to comprehend physics but his low mathematical ability. Once we identified this gap we switched our focus to improving his mathematical abilities and the problem with physics went away.

How do you see KPE in your journey?

> In Uganda, national assessments of early literacy and numeracy skills acquisition reveal that seven out of ten children spend up to three years in school without ever learning to read or write, and about the same number are innumerate as well. I thought one of the main problems might be how the children learn to read, especially those whose first language was not English – the official schooling language.
>
> In recent years the phonics approach to teaching how to read had been introduced into schools countrywide [163]. However, my observations in some rural schools was that the instructional materials to support this new way of teaching had not been widely distributed. To make matters worse, the materials available to buy in bookstores were not well adapted to the local context because they often included cases that were not locally relevant. For example, depictions of children riding "ponies" or picking "apples" or playing in the "woods." *In Uganda we do not have any of those naturally.* Not to mention that these children in the teaching material had unfamiliar names like Tiffany and Aiden, which to most Ugandan children were not recognizable as the names of people, and so did not help their learning along, see Figure 6.2.
>
> My idea, then, was to recreate some of these simple instructional materials by replacing the inappropriate examples with more contextually appropriate examples. One example is referring to local fruit (e. g., jackfruit), with the corresponding local names (e. g., Kato and Okello) and the local landscape (e. g., anthills and maize gardens).
>
> When I shared my idea with teachers in these poor rural and urban schools I found out that actually the problem that the teachers faced more was that they did not know how the phonics sounded, and so could not teach using them. Even before having contextually relevant instruc-

Figure 6.2: Children eating the famous jackfruit (courtesy of https://ileftmyheartinuganda.com) and Connie reading the book written by Ngũgĩ wa Thiong'o's: Decolonising the Mind.

tional books they would benefit from audios which helped them learn how to pronounce the phonics – something that I took for granted – and then my adapted instructional materials would be useful after that.

Since then I have understood that people experience the same problem in different ways, and that in creating a solution one's first idea will most likely not fit the end-user that well. It is therefore important to validate the nature of the problem before one goes too far with developing the solution.

6.2.3 Swab sensor

Richard Novak
Co-founder & CEO of Unravel Biosciences Richard Novak
Richard is a bioengineer and entrepreneur dedicated to solving unanswerable questions through technology innovation. Prior to Unravel Bio, he was a Lead Engineer at the Wyss Institute for Biologically Inspired Engineering at Harvard University, where he managed fast-paced programs in drug discovery, advanced disease models, human organ chips and integrated automation and sensor systems. He has over 15 years of experience in microfluidic system development and applications in the therapeutic space. He is a founder and president of the nonprofit Future Scientist and a founder and director of the sample collection automation company Rhinostics Inc.

We make clinical disease diagnostics accessible to everyone

Rhinostics was founded during the COVID-19 pandemic to address the immediate need to bring diagnostics to everyone by making it widely available and fast. A limiting factor, aside from the lack of sample collection swabs early in the pandemic, was the throughput in the clinical diagnostics labs (CLIA labs). The biggest bottleneck was making collected samples compatible with automation equipment, with labs hiring sometimes hundreds of temporary workers to ... remove swabs!

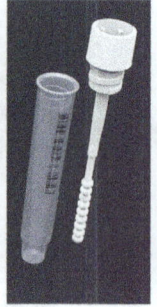

Figure 6.3: RHINOstic automation-compatible swab and barcoded tube that interfaces with automated laboratory workflows to scale up diagnostic capabilities. A batch of nasopharyngeal swabs manufactured from one-shot injection molded polypropylene. Richard Novak (right) together with Future Scientist project coordinator John Michael Coatney and key partner, Anibal Villacres meeting in Anibal's town of Nombre de Dios, Panama.

By combining a novel swab head design with a threaded cap compatible with decapping robots, we solved two problems at once: (1) we manufactured swabs using highly scalable one-shot injection molding, and (2) we solved the problem of automation compatibility for a dramatically faster and less expensive process, see Figure 6.3.

Understanding the problem around swab shortages during COVID, especially the really tough job of CLIA lab workers dealing with the risk and tedium of removing bits of plastic from tubes, led to this solution. Understanding came through empathy from speaking with the people in the trenches. From that empathy, we built our knowledge of swab collection, clinical diagnostic needs and regulatory processes, which built on our existing knowledge of microfabrication and diagnostic assay design. Persuasion played a strong role at the middle and end stages of development, where we had to convince overloaded clinicians to part with a few minutes of their time to provide design feedback and eventually run clinical trials.

As the company grew, we built a team of professional persuaders: a sales team, whom we selected specifically for their ability to empathize with potential customers. The goal has been to solve a problem, and Rhinostics was built around empathy to enable it.

Was "empathy" ever taught or discussed as a relevant skill you needed, for either work or your personal life?

> I believe that in practice, one's education takes place largely through experience and periodic mentorship or guidance. During my personal upbringing (family, friends, other nonprofessional activities), before, during and after university years, the concept of empathy and its importance was instilled in me. As an immigrant to the USA with my family, I grew up on the receiving end of highly empathetic people who supported us, in turn enabling me to pay it forward and appreciate the range of challenges faced by others.

The one time I was formally taught empathy was as part of a design engineering course, where understanding a problem takes place through understanding the people who experience the problem. I was taught there that empathy brings the designer closer to the problem by experiencing the problems more fully. IDEO has formalized this concept in a Human-Centered Design Toolkit (https://www.ideo.com/post/design-kit) that, while aimed at developing regions, can find application in any situation.

How do you see KPE in your journey?

To me, empathy permeates the entire process, start to finish. Yes, one must have some knowledge before embarking on a solution, but really this can take place after an empathetic characterization of a particular problem. Many entrepreneurs are gifted with the ability to identify real problems and start companies off of that, not off of a solution, which they develop later by bringing in relevant knowledge through hiring talented scientists and engineers. The gap between initial empathy and the ability to solve a problem is bridged with persuasiveness: Persuasiveness to secure initial funding and take a risk on the ability to empathize with potential future customers, the persuasiveness to recruit the talent with the appropriate knowledge and the persuasiveness to then bring the solution to the market that needs it. Of course there are many other permutations equally possible. My current company, Unravel Biosciences, started as a research project with a heavy dose of knowledge around ways to work with the complexity of biology to develop effective therapeutics. We didn't accelerate until we identified Rett syndrome, a severe neurodevelopmental genetic disorder, as our initial disease to target. How did we do that? Empathy toward a couple with a daughter with this disease. The discussion with them opened my eyes, not only to the clear medical challenge but also to the failures of traditional pharmaceutical development. Our founding team brought together (through persuasion) tremendously skilled scientists and engineers to tackle this disease and others, leading to the formation of Unravel Bio as a way to accelerate therapeutics to treat patients based on the full extent of their needs.

IF example

The first factors we need to focus on are "collected mass," as new swabs collect slightly less material, and "mass release efficiency," because they release nearly everything, nearly instantly, unlike absorbent swabs. Then, for the "time per swab removed," automated decapping accelerates the physical process of removing the swab vs manual removal. We also have the "sample accessioning time," where integrated barcodes on the vial allow for automated scanning to accession samples into the sample management system. Lastly, the number of people needed to carry out the tests is reduced due to the ability to use automation throughout the laboratory process, see Table 6.2.

6.2.4 Chemistry is in the air

Stafford W. Sheehan
Co-founder & CEO of Air Company
Dr Stafford W. Sheehan is the Chief Technology Officer and Co-founder of Air Company. The company uses captured CO_2, H_2O and renewable electricity to produce alcohols, making a carbon-negative product that acts as a method to sequester carbon dioxide in the chemical bonds of a material that

Table 6.2: Comparison between conventional diagnostics (Before) and the method developed by Rhinostics (After).

Case	Factor	Before	After	d	IF
Rhinostics	Collected mass [relative]	1	0.75	−1	0.75
	Mass release efficiency [%]	20	99	−1	4.95
	Time per swab removed [min]	10	3	1	3.33
	Sample accessioning time [min]	2	0.1	1	20
	Staff to process 30k samples/day	75	5	1	15
				IF_{total} =	16.40

displaces an equivalent fossil fuel. Air Company has scaled its technology and business to over 30 employees in New York and New Jersey in the USA. Stafford received his PhD from Yale University in 2016, studying chemical methods for artificial photosynthesis and related systems, and for the last five years he has been developing industrial technology for CO_2 utilization.

Developing CO_2 conversion technology to replace fossil fuels

Solving climate change is at the core of what Air Company does. The business was founded to utilize carbon dioxide in ways that are net-negative, or net-neutral in terms of greenhouse gas intensity in the atmosphere. All of Air Company's products, both consumer and commercial, serve to advance and scale technology with the eventual goal to produce fuel using carbon from the air, water, and renewable electricity, to end the need to drill for fossil fuels, see Figure 6.4.

Figure 6.4: Air Company founders, Gregory Constantine (left) and Stafford Sheehan. To the right, some impressions of the plant and one product they make: vodka. The *IF* example can be seen in Table 6.3.

Table 6.3: Conventional corn-based ethanol (Before) with the CO_2-based method developed by Air Company (After).

Case	Factor	Before	After	d	IF
Air Company	GHC emissions [gCO_2e/MJ]	74	−17	1	1.23
	Water Cost [$/kg EtOH]	0.18	0.002	1	90.0
	Land required to produce All ethanol [acres/year]	50.9 M	3.100	1	16.4 M
					$IF_{total} = 1.8M$

Was "empathy" ever taught or discussed as a relevant skill you needed, for either work or your personal life?

Whether explicitly acknowledged or not, pursuing a career in the climate space requires empathy as a relevant skill. Responses to climate change are inherently driven by empathy – namely, the desire to mitigate a crisis with disproportionately heavy impacts on those least responsible for it. Simultaneously, climate change exposes the fundamental flaw with the dominant iteration of recent capitalism – its prioritization of profit accumulation over empathy, as exemplified by the business practices of the fossil fuel industry. Given the influence this economic structure has and the values it imparts on us, it is no surprise that empathy is not regularly taught as a relevant skill. In tackling climate change, businesses must consider not only quarterly results, but also the impacts they are having on fellow humans. In this way, climate change presents an opportunity to both learn and unlearn – to reimagine our current economic model to be more empathetic, humane, and equitable. Such an opportunity presented itself in the first year that we launched our products (2020), when the COVID-19 pandemic outbroke and empathy guided us to pivot our ethanol production for use in hand sanitizer. To date, we have donated over 16,000 bottles of sanitizer produced from atmospheric CO_2 and are exploring further philanthropic deployments of our technology.

How do you see KPE in your journey?

As a company whose mission is to mitigate climate change, E underpins everything the company does. As R&D engineers, we gather and create K that informs where and how we can best act in line with E. Without rigorous research and nuanced K, we cannot determine the best ways to uphold the pursuit of E. P is critical to the success of this mission, as consumer, investors, and policymakers must be willing to engage with nascent innovations that can reduce CO_2 concentrations and replace fossil fuel usage. This P is bolstered by the K we acquire through our research and relies on appeals to the E that gives us purpose. Thus, K, P, and E are all interwoven and fundamental to our journey.

IF example

The *IF* calculations on the comparison between conventional corn-based ethanol with the CO_2-based method developed by Air Company are shown in Table 6.3. The values have been taken from [143, 61, 108].

6.2.5 The challenges are there, but can we see them?

Akash Raman
Akash is a PhD student at the Mesoscale Chemical Systems group in the University of Twente. Under the supervision of David Fernandez Rivas and Han Gardeniers. Akash is working on improving gas-evolving electrodes by optimizing mass transport. His PhD focuses on the influence of bubbles on the electrochemical performance of electrolyzers. He has worked on numerous topics, including water treatment, electrochemistry and flow chemistry. He is interested in developing innovative solutions to improve access to clean water and energy.

The challenges are there, but can we see them?

Challenges and opportunities are two sides of the same coin. At the root of every challenge lies a problem waiting to be solved by a scientific (not necessarily STEM) mind. Of course, not every challenge is an entrepreneurial opportunity – nor should it be.

Naturally, the zeroth step in any entrepreneurial process is identifying an opportunity. Humans are egocentric by design and we tend to prioritize our immediate needs over many nobler causes. The effect of this egocentric (or unempathetic) approach is far-reaching. It biases the entrepreneurial process towards the so-called First World problems. Consider the problems that were tackled in the last century. Nearly all were purely technical. Can we send a man to the moon? Can we make cars affordable? Can we make aircraft fly farther, faster? This scenario is slowly but surely changing. Problem solving today must invariably take into account the socio-economic impact. Not only do the most pressing problems we currently face have enormous social effects (such as climate change disproportionately affecting the poor), but markets are also beginning to adapt to an increased appetite for socially conscious initiatives.

This is where empathy plays a crucial role. A conscious attempt at empathy (which surely most of us are capable of) can help sway the entrepreneurial process toward some of the most pressing challenges of our time without overtly stymying general capitalistic innovation.

Was "empathy" ever taught or discussed as a relevant skill you needed, for either work or your personal life?

Empathy was a natural part of my upbringing. It is impossible to grow up in India without witnessing the everyday struggle of millions of your compatriots. Although, thankfully, I did not have to face such struggles for my basic needs, I was very aware of the possibilities and consequences. That being said, it is easy to "forget" those experiences and lose touch with one's empathetic side.

A specific wake-up call arrived when I received the opportunity to work on a project led by the Indian Institute of Technology Madras and the Vivekananda Kendra (a social organization). The project involved traveling all over the small island town of Rameshwaram and checking the status of scores of biogas generators installed by the program a few years ago, see Figure 6.5.

While I joined the project as a chemical engineer (in the making), I quickly realized that what was more important was measuring the impact of the plants. One of the aims of the program

Figure 6.5: (Left) Akash (leftmost) interacting with end-users of the household biogas plants at Rameshwaram. (Right) The household biogas plant (black cylinder) enables users to sustainably fertilize small crop fields.

was to replace wood-burning stoves in households with gas stoves. The biogas generators could generate gas from household waste and also help with waste disposal and sanitation.

All this, while providing organic fertilizer (the runoff from the generator is nutrient-rich) to a largely agrarian community. The relatively simple system – with few moving parts and little to no maintenance – opened my eyes to the value of innovation driven by empathy.

How do you see KPE in your journey?

As a PhD candidate (at the time of writing), the role of knowledge is clear and, quite frankly, fundamental to my line of work. I must however admit that I've overemphasized the K aspect thus far. I haven't fully appreciated the impact that the other two components have had on my career.

Empathy, in particular, has played a pivotal role in bringing me this far. The professor who recommended me for the project in Rameshwaram did so because he knew I could offer more than an engineering perspective to the project. While on the island, it dawned upon me that the primary resource that was deficient on the island was energy. This realization, which stems from empathy, continues to motivate me to pursue a career in clean energy technologies. Most people I compete with for academic positions have good, if not better, academic credentials – the K factor doesn't play a determining role. It helps if you can demonstrate P and E. This goes to show that the KPE framework is extensible beyond entrepreneurship – it can provide tangible benefits to one's career while still making meaningful contributions to society.

6.2.6 Flying droplets

Tom Kamperman
Co-founder & CTO of IamFluidics BV, postdoctoral researcher University of Twente

Tom is an entrepreneurial biomedical engineer with a strong background in microfluidics, tissue engineering and enabling microtechnologies. He has over eight years of experience with the development of microfluidic systems for life science applications and in particular microencapsulation. He has been working at various internationally recognized institutes including the Max-Planck Institute in Muenster (Germany), the MESA+ Institute for Nanotechnology (Netherlands), the Technical Medical Centre (Netherlands) and the Division of Engineering in Medicine at Harvard Medical School (USA). Tom received the "Overijssel PhD award," the "NBTE PhD thesis award" and the "DutchBiophysics BIOPM thesis award" for his PhD thesis titled "Microgel Technology to Advance Modular Tissue Engineering." In 2019, he was elected "European Innovator Under 35" by MIT Technology Review. Tom is co-inventor of the "in-air microfluidics" technology and co-founder of the company IamFluidics BV, where he is responsible for technical management and product development. Tom is currently also affiliated as a postdoctoral researcher at the University of Twente, where he is working on single cell encapsulation to protect cells during 3D bioprinting.

We revolutionize microparticle production

Microparticles are everywhere. Products in cosmetics, nutrition and pharmaceutical markets are nowadays based on or contain microparticles. For example, encapsulation into microparticles enables masking of bad tasting nutrients, controlled delivery of medicines or long-lasting release of flavors and fragrances. However, the industry is currently facing a limiting trade-off between the quantity, quality and sustainability of microparticle production. IamFluidics developed and commercializes a patented technology called "in-air microfluidics" (IAMF) to overcome this paradox, see Figure 6.6. IAMF technology integrates printing technology with microfluidics, thereby enabling the industrial-scale production of high-quality and sustainable microparticles at rates 1000× faster than conventional high-quality production methods while achieving micron-resolution with minimal energy losses.

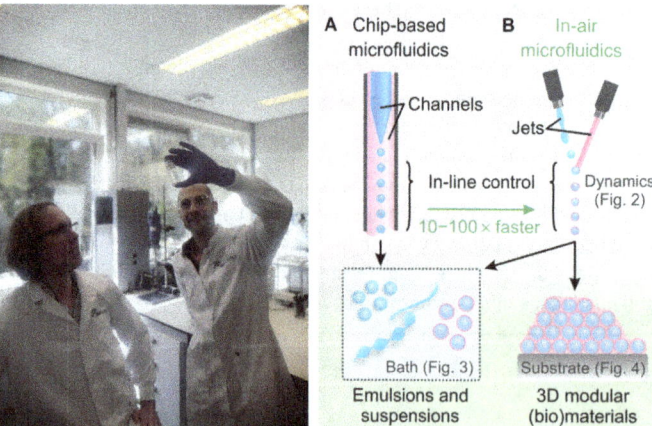

Figure 6.6: Tom Kamperman and co-founder Claas Willem Visser in the lab. Image depicting the backbone of IAMF technology reproduced from [175].

Was "empathy" ever taught or discussed as a relevant skill you needed, for either work or your personal life?

> Yes, during my personal upbringing and during my professional path. My parents always taught me to respect and empathize with other people. That mindset also helped me to work with different people and teams in a multidisciplinary professional environment later on.

How do you see KPE in your journey?

> Until today, I never considered or realized that "empathy" is part of this equation or development path. It is quite an eye-opener actually.
> I think these profiles depend on the role of each person in the company. As CTO/co-founder, I am naturally more focused on the K. I think it is fair to say that the general message and balance of KPE make sense, but that there are certainly differences depending on the moment or timing of entering the company, your personal character and your professional job description. Importantly, just showing this KPE to people might help with a certain realization if they are completely off-balance.
> Empathy still feels a bit uncomfortable compared to the other two (K & P). It seems to be a completely different category. That said, it does make sense to include it or present it like this. Still, I'm glad that I've seen it in this perspective; quite helpful!

IF example

IamFluidics uses its IAMF technology to revolutionize microparticle production. IAMF is a scalable technology that lends itself to large-volume production. Compared to conventional chip-based microfluidics, IAMF provides up to 1000× faster per-nozzle flow rates and is not hampered by channel clogging. Its open, i.e., "in-air," nature enables swift production scale-up and implementation into environments that are incompatible with commonly used closed systems. Furthermore, IAMF eliminates the need for high voltages or rotating equipment, which are required in conventional jet-based methods currently used for production of high-quality microparticles. IAMF is therefore safer and compatible with a vastly wider range of materials, including the processing of flammable and sensitive liquids, see Table 6.4.

Table 6.4: Compared to conventional chip-based microfluidics, IAMF technology offers several advantages such as better environmental safety by reducing solvents and surfactants. Here are some advantages included in an *IF* analysis.

Case	Factor	Before	After	d	IF
IAMfluidics	Versatility [relative]	1	2	−1	2
	Collected mass [relative]	1	100	−1	100
				IF_{total} =	200

7 Epilogue: Last considerations

Horace, a Roman poet who lived from 65 to 8 BC, once said that a book can be a "monument more lasting than bronze." He apparently meant that after writing a book he would not wholly die.

In my birth country, Cuba, it is typically said that you have to plant a tree, write a book, and have children before you die. As you can see in Figure 7.1, I got a tree together with Bram Verhaagen, my friend and co-founder of BuBclean. I hope not to die anytime soon, but getting to this stage in life brings me great joy.

Figure 7.1: (Left) Bram Verhaagen standing next to our tree, which we received in November 2013 as one of the 75 innovative companies that symbolically received a tree at the Lane of Innovation in Enschede, The Netherlands. The trees represent sustainable growth of innovative companies in and from Twente. (Right) Bram and I proudly holding the certificate of the tree, which luckily we were not supposed to water or maintain afterwards.

I never really thought I would write a textbook, and when the idea started itching, I read several blogs and asked colleagues who had done it, journalists, etc. The overall message I got is that you should write it as long as you believe you have something useful to say. But how do you know when what you have in your head can be of relevance to others?

My "a-ha" moment when I decided to go ahead came while watching the documentary video based on "Letter to you," an album by Bruce Springsteen [123]:

I'm in the middle of a 45 year conversation
With these men and women I am surrounded by, and with some of you (…)
I tried to make that conversation, essential, fun, and entertaining.
I started playing the guitar
Because I was looking for someone to speak to
And correspond with
(…) All I know is that after all this time
I still feel that burning need to communicate

It's there when I wake every morning
It walks alongside of me throughout the day
And it's there when I go to sleep each night

Over the past 50 years it's never once ceased …
Owing to what I don't really know
Is it loneliness, hunger, ego, ambition, desire, a need to be felt and heard, recognized, …?

All of the above
All I know is that is one of the most consistent impulses in my life
As reliable as the rhythmic beating of my own heart

Is my need to talk to you

♪ ♩ Bruce Springsteen (American singer-songwriter) – Letter to you, 2020 [123].

I am a bit younger than "The Boss," but I also felt the urge to channel this energy and drive to communicate with my students and colleagues, and also those of you that I have not had the chance to reach so far.

I know it is hard to solve problems from a student's room or in social distancing settings. But if we have "our brain" and some degree of health, there is hope that we can try to solve societal-relevant problems. This book comprises plenty of communication and collaborative tools and tips at your disposal, and I really hope it can help you.

I trust that after reading this book – or at least some parts – you might have answers to the questions that may have popped up in your head after reading its title.

- What is an empathic entrepreneurial engineer?
- Which is my missing or less developed ingredient in KPE?
- When, how and where can I start solving problems?

My last suggestion to you is to reach out and join existing research groups' activities at any educational or research institution or company. It is a big hype nowadays to have citizen-science projects, also known as community science, civic science or volunteer monitoring: research conducted, in whole or in part, by nonprofessional scientists.

Alternatively, or in parallel, you can also join broader or global-scale initiatives to solve urgent problems. The list is long: global warming, plastic contamination, finding a cure for cancer, ...

If you need any sort of guidance in this context, your teachers and I are happy to help you! Please, check https://empathic-engineering.com, for updates.

In the end, we should always focus on learning. There is a tendency to focus on the final result of a project, a product or success in any measurable form. However, we are humans, living creatures continuously making mistakes, and hopefully learning to adapt.

We should all be given the chance to learn and grow as individuals. I am even learning as I finish this book, which I am sure could have even more examples or a sharper approach if I had taken more time – but then the longer this book would be in my hands, the faster it would lose relevance and it would take longer to reach your hands. I will do my best to collect input and update this book at www.empathic-engineering.com.

I cannot claim to have found all the best tools to help you be more empathic, persuasive or entrepreneurial. I would not even dare to imply I gained all the knowledge I will need for my future professional or personal trajectory. But I am happy to end this book wishing you success and hoping you learned new things by reading it.

Thank you, once more, for the time and please share your story while mixing your KPE ingredients!

A Answer to case study questions

A.1 Knowledge answer

This answer corresponds to the question in Section 2.3.3. Let us use as real-life example my PhD project, financed by the Dutch Technology Foundation STW, with the title "Efficient Sonochemical Microreactors." The word "sonochemical" stands for the use of ultrasound to generate physical and chemical effects. The research goal was to improve the energy efficiency of existing sonochemical reactors, by at least one order of magnitude.

My main objective was to design, develop and test energy-efficient sonochemical microreactors. There was the option to miniaturize existing reactors or invent new ones. I chose the latter, which allowed me to gain control over the places of bubble formation and study with unprecedented detail their collapse (cavitation process).

We can agree that the whole description about bubbles rising from the surface of a glass containing water with saturated gas sounds irrelevant at first sight. Since it is an accepted fact by most people, you could say that this falls under *information*. But if we use that information to improve the energy efficiency of sonochemical reactors, then it becomes *knowledge*, as you will see in what follows.

Most of the applications of ultrasonics sonochemistry have been made in the laboratory or small batch reactors. In contrast, industrial large-scale sonochemical applications have been limited by the inefficiency of converting electrical energy into a chemical product. The conversion of electrical power is done with an ultrasonic transducer. During the whole process there are heat and energy losses. Interestingly, bubbles in a cluster can shield each other, diminishing the potential energy of collapse. This is already a hint that 'more bubbles' it is not a readily good solution.

When I started my project, I first looked at different ways in use to measure efficiency and decided to define the sonochemical yield of a given process as the ratio of measured chemical effect to the energy injected to the system. This formula gives typical values on the order of 10^{-6} (see equation (3.1)). Most of the energy loss is known to be associated with bubble cavitation phenomena, because the electrical transfer can not be easily optimized.

A bubble is a gas cavity surrounded by liquid. Each oscillating bubble, growing and collapsing in a fast and adiabatic way, is in fact an individual high-parameter reactor. The contents of the bubble will be heated and pressurized in a dramatic way, leading to an energy-focusing effect. If you want to know more about this fascinating phenomenon, you can read my PhD thesis [88].

I spent roughly one year attempting to gain control over the generation or nucleation of bubbles. My thought was that after such control would be achieved, I would be able to set bubbles into cavitation with well-specified conditions. The first attempt was by using a laser to heat a small volume of liquid in a microchannel [87], followed by using a surface with small bubbles stabilized in artificial crevices or defects exposed to shockwaves (based on previous work [57]); but none yielded any measurable chemical effect.

In the end, my team of collaborators and I succeeded with a reactor configuration having continuous ultrasound irradiation. We used the same smooth surfaces with artificial crevices of the shockwave setup explained in page 50 and Figure 3.7. The way it works is similar to the carbonated water in the glass question, in which the bubbles emerge from specific places at the inner surface of the glass. We then fabricated crevices as artificial defects in silicon substrate surfaces[1] – which we called pits – to stabilize a bubble when submerging the substrate in a liquid.

When applying continuous ultrasound to a small container with the substrate covered by water, I saw at first, at relatively low pressure amplitudes, interesting effects. Since my experiment was not superclean, particles dissolved in the liquid started moving around, and some of these particles accumulated on top of the oscillating bubble in the pit. These particles were ejected every now and then in a way resembling the eruption of a volcano. This was cool, but not what I was looking for.

As I increased the pressure amplitude, even more interesting things took place: the appearance of microbubbles that traveled in strange trajectories that resembled some fireworks or scenes of "Star Wars" like in Figure A.1, left image. To make a long story short, in the end we managed to increase the energy efficiency by an order of magnitude.

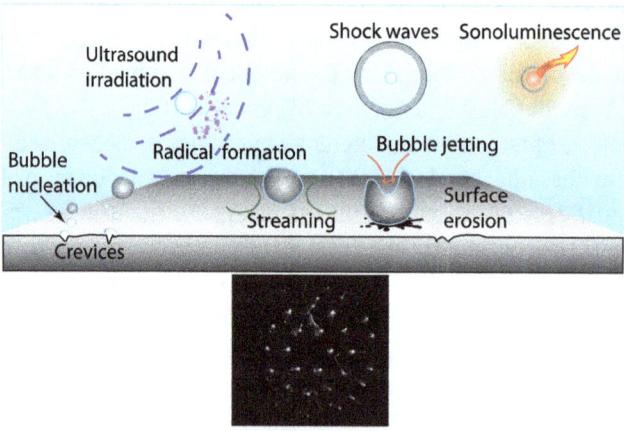

Figure A.1: (Top) Schematic representation of the main effects of cavitation induced by ultrasound irradiation (sonication). Reproduced from [91] with permission from the Royal Society of Chemistry. (Bottom) Bubbles nucleating from a 42-pit array arranged in a concentric way. The white regions are the nucleated bubbles contrasting with the dark background [88].

---

1 Using the same technology that allows fabricating semiconductors and microelectronic components.

A.2 Persuasiveness answer

This answer corresponds to the question in Section 2.3.4. There is no "right or wrong" way to answer this one (as with the other questions in this book). It is all a matter of your personal characteristics and your ability to match the situation you are facing. Most people would say, use "common sense," but that is too vague and what is common today may make no sense in two years or in a different country.

Before diving into the actual answer, I want to give a short anecdote. During my first meeting with a very famous innovator, I started describing the potential for market utilization of my needle-free injection idea. In particular, tattooing was and is a very popular activity, and I believed that we could avoid many of the health and environmental problems caused by this old technology based on needles. At the same time, its market entry would have the potential to make a good business case.

While I was talking I realized that the focus of my interlocutor waned and he started looking outside the window. I instantly spotted that something was off, and recalled a similar situation earlier that year, in a meeting with another brilliant scientist who directly stopped me and asked: "why should we improve the way tattoos are made?"

Of course I had already answered that same question myself, and came to the conclusion that:
- People were tattooing themselves for millennia and that would not change overnight.
- Similar to the problem with drug addicts, who use new needles regularly, to avoid transmission of infectious disease and other complications, I told to myself (and my collaborators): if we can provide a safer and less contaminating way to inject, or tattoo, then we are going to solve a real societal problem – whether we agree with the esthetics of successful music or sports stars or not.
- Not all tattoos are for fun; several tattoos are made for medical purposes, to camouflage wounds, to cover areas with severe hair loss or to redraw the nipples of patients after breast removal resulting from cancer.
- And if the above was not enough, if the same method to inject could be used to treat diabetes or vaccinate . . . why shouldn't we try it?

It may not come as a surprise that when I talked to people with private capital, they were more interested in hearing the cosmetics or tattoo-for-fun story because there is/was a shorter path to get investments back. Most people are familiar with the fact that medical developments require lengthy and expensive medical approvals, which often do not end well.

Thus, I chose to give you here my answer roughly describing what I did when preparing for defending one of my most beautiful and challenging projects, and it was to persuade a scientific jury about my research proposal:

- ☐ 3. The most commercially appealing (though controversial to some)
- ☐ 2. Use a joke to set the tone
- ☐ 4. Why you are the best candidate to get the money
- ☒ 1. Start with thanking for the opportunity and the biggest scientific challenge

No matter the instance where you get the opportunity to stand in front of other people to talk, please, always start thanking your audience for the chance to be listened to.[2] Moreover, it is customary to thank the organizers of the event or the team that worked together at the end of a presentation during a conference. On online platforms and in blogs or videos, this may be less needed because time is very valuable on those platforms, and you can always put it elsewhere.

ⓘ "The European Research Council (ERC) supports frontier research, cross disciplinary proposals and pioneering ideas in new and emerging fields which introduce unconventional and innovative approaches. The ERC's mission is to encourage the highest quality research in Europe through competitive funding and to support investigator-driven frontier research across all fields of research, on the basis of scientific excellence."

European Commission
(https://ec.europa.eu/programmes/horizon2020/en/h2020-section/european-research-council).

I applied for the ERC Starting Grant, meant for young, early-career researchers (2–7 years after their PhD), and I showed an emblematic passage of the popular TV show Star Trek on the first page of the research proposal text (see Figure A.2). I also use regularly the video during the presentations of my project in front of panel members and wide audiences. Logically, I took a risk with this rare way of using elements not purely traditionally associated with a serious scientific project. But I knew this may resonate with the audience, nerds or lovers of science fiction, like I am.

The idea ended up working quite well. Playing such video or showing the image helped me to connect with the readers of my scientific proposal and relax a bit the tension inside the room where I was trying to convince the jury about the quality and innovative approach of my way of working. You can see more about this project on www.bubble-gun.eu. I like to believe that my originality in presenting a very complex project and the answers I gave, together with the review report given by experts, helped me persuade the jury to grant the funds for the project.

To avoid uncomfortable situations in the particular setting of the ERC proposal, I typically start with the more medical potential applications, such as insulin injection or vaccinations. Only if the topic of commercialization pops out (which it regularly does), then I explain my vision and the commercial appeal of injecting without needles for cosmetic applications, such as recreational tattooing.

---

[2] See the very first page of the Preface of this book, page XVII.

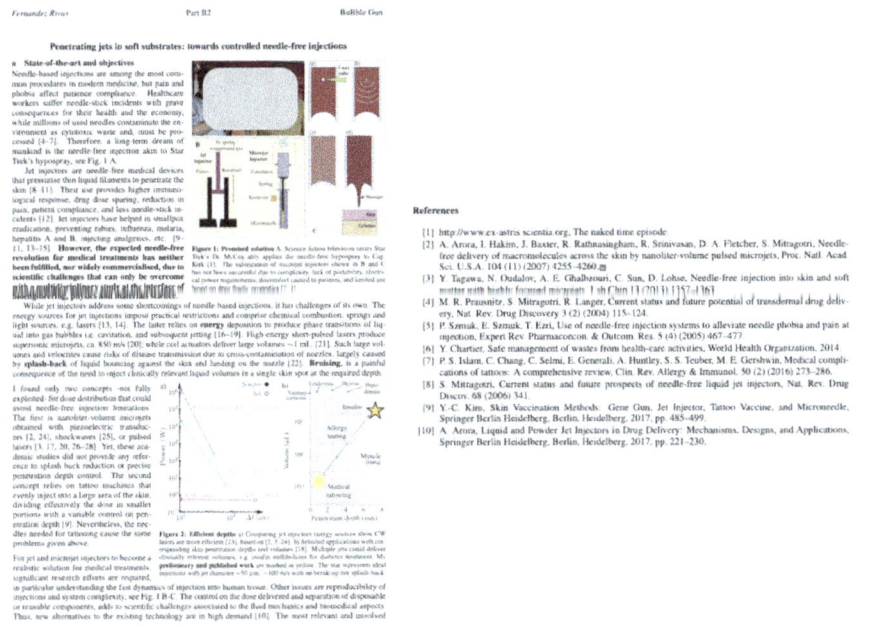

Figure A.2: First page of the content described in the proposal for BuBble Gun: Penetrating jets in soft substrates: towards controlled needle-free injections. Notice the use of an image typically not associated with scientific settings. In trying to connect or empathize with the jury members, I used a reference to Star Trek, a science fiction TV show that predicted several inventions. One of them, needle-free injections, appeared in several episodes (ViacomCBS/Paramount Pictures). The figures on the proposal page have been used with permission. Bottom left: Modified from [50]: Copyright (2007) National Academy of Sciences, USA.

It is important to note that a typical rate of success for this sort of projects is as low as 10–12 %. It is very hard to predict what can work best, but persuading people to give their time, or money, no matter if it is theirs or they are managing it for someone else, is a skill that should not be underestimated.

I believe it is possible to persuade other people with a warmer tone by using humor. My final answer or advice would be, make an authentic effort to show what your ambition and plans are. I chose for humor in this example because I was confident that I could also answer the most complicated questions, not only technical, but also ethical. For example, I was asked by one jury member during the questions & answers session if the same technology could be used for injecting poison or other unethical activities. What do you think was my answer?

A.3 Empathy answer

A possible answer follows to the assignment given in Section 2.3.5. In my courses, I provide this as an assignment to the students, who prepare their document as an

A Answer to case study questions

"Empathy report" to be reviewed by another student (or group). Here, in Figures A.3 and A.4, I provide a couple of these assignments.

A.3.1 Answer 1

Note these students went beyond the requested page limit.

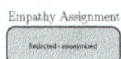

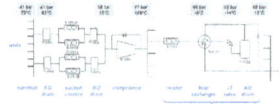

Figure A.3: Answer 1. The image in the Problem description is included with permission from the Dutch National office for Cultural Heritage (*Rijksdienst voor het Cultureel Erfgoed*) (CC BY SA 4.0).

A.3.2 Answer 2

Empathy assignment – Innovating Reactor Systems (Redacted - anonymised)

The Flint water crisis

Introduction

Flint is A city in the USA in the state of Michigan. It has been a hot topic in the last few years because of an ongoing drinking water crisis. In 2014 the city switched from it's regular source of drinking water, Lake Hurron, to the Flint river to save money. This caused multiple problems, first of all this water source contained more harmful bacteria such as Escherichia coli and also contained harmful elements of disinfectants that were used to remove the bacteria. However, most importantly this new water source was significantly more corrosive than the old water source and thus corroded the metal pipes and leached metal particles into the drinking water distribution network. Because a high percentage of the waterpipes in the city were outdated and made of lead this caused a significant increase in the amount of lead in the water.

Lead is a neurotoxin and can cause a decrease in intelligence and future life achievements when kids are exposed to large amounts, such as in Flint. A scientific study showed an increase in blood lead levels after the switch of drinking water source in 2014[1]. This was especially prevalent in poorer neighborhoods in Flint. As can be seen in figure 1.

Initially the state denied that the quality of the water was insufficient. Therefore the residents of Flint were exposed to water with increased lead levels for 18 months. When it eventually did come to light, free bottled water was supplied to the residents of Flint. Unfortunately due to the cost no more free bottled water is handed out.

Next a point of view from a resident will be told based on several real experiences in Flint[2].

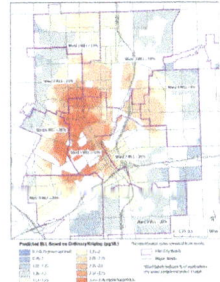

Figure 1. Blood lead levels in Flint Michigan.

Point of view from residents

Alexandra (fictional) is a single mother living in Flint Michigan. She and her children have been exposed to significant amounts of lead because the city and state officials cut corners to save money. These officials do not live in Flint themselves and don't have to use this contaminated water. Alexandre feels cheated by these officials. The city had initially provided free bottled water and free filters for the tap. However, no measure is taken for people to wash themselves and therefore Alexandra still has to shower in this contaminated water. From this she and her children have developed a horrible rash. The medicine against this rash are too expensive for her and other people in the poorer neighborhoods of Flint. Therefore she has to choose between a rash and not cleaning herself.

Solution

As told above, initially free bottled water was supplied to the residents of Flint to replace the contaminated drinking water supply. However, due to the cost of this measure this could not be continued indefinitely.

Therefore it was chosen to distribute free water filters to the residents of Flint. These so called 'point of use'(POU) filters can be attached to the tap to remove lead from the water.

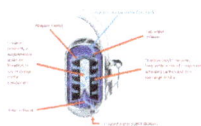

Figure 2. A typical POU filter.

A scientific study showed that these POU filters decrease the lead content well below the bottled water standard[3]. The highest concentration measured was 2.9μg/L, which was well below the standard, and over 97% was lower than 0.5μg/L.

These filters can not be applied for the shower because hot water can actually reverse the chemical adsorption process and therefore lead cannot be filtered out. And cold showers are not an option in Michigan where it can get very cold in the winter.

Therefore I would suggest developing a shower head for people in flint that consists of two parts. First a part where cold water is filtered using a filter such as described above. And a second compartment with a heating element to make the showers more comfortable.

The filter would be slightly less effective to increase the throughput. A slightly higher amount of lead will remain in the water, but since the shower water does not have to be consumed this is not as much of a problem.

The current shower head filters that are available cannot remove the lead and only removes chlorides. However, it is not known which material exactly causes these rashes in Flint and therefore a more complete filter such as suggested would be more effective as it not only removes the chlorides.

This suggested shower head will hypothetically improve Alexandra and her kids quality of life as it will help against these rashes.

Discussion

The POU filters are not always effective in real life, because often people do not know how to install them properly or when to replace the filters[4]

literature

1. Hanna-Attisha, M., et al., *Elevated Blood Lead Levels in Children Associated With the Flint Drinking Water Crisis: A Spatial Analysis of Risk and Public Health Response.* American Journal of Public Health, 2016. **106**(2): p. 283-290.
2. Levin, S. *Still standing: Flint residents tell their stories about living with poisoned water.* 2020; Available from: https://www.mlive.com/news/page/still_standing_flint_residents.html.
3. Bosscher, V., et al., *POU water filters effectively reduce lead in drinking water: a demonstration field study in flint, Michigan.* Journal of Environmental Science and Health, Part A, 2019. **54**(5): p. 484-493.
4. Guha, A. *State Water Filters Prove Lacking in Flint, a City 'Full of Forgotten People'.* 2018; Available from: https://www.innewsgroup.com/article/2018/08/16/state-water-filters-prove-lacking-in-flint-a-city-full-of-forgotten-people/.

Figure A.4: Answer given by students.

A.4 IF cleaning answer

Answer to Exercise X1, same page where Table 3.2 is, just below. Can you calculate the *IF* value? Please, find the calculated values in Table A.1.

Table A.1: The superiority of using the Cavitation Intensifying Bags (CIBs) in ultrasonic cleaning of jewelry parts, evidenced by innovation/intensification factor (*IF*) values larger than one.

Case	Factor	Normal	CIB	d	IF
Jeweler	Time [min]	10	2.5	1	
	Volume [L]	3	0.05	1	240

This example is a bit simple, but you may be surprised how relevant it is in commercial settings. Same as most people are not aware of how much money it costs to do a laundry wash at home. For example, we may know the cost of electriciy usage, but it is a bit more difficult to sum all separate costs, say for example, water, detergent, time to put the dirty clothes in and out of a washing machine, etc. This calculation helped us to make a point with the customers of the bubble bags, see http://www.bubclean.nl/bubble-bags-2/.

A.5 Examples used in lectures to teach *IF*

A teasing question I love and have used in many lectures is *"Who is the best president?"*, while looking at a picture with a line-up of different presidents (from the past and current).

Without entering a political discussion or sociopsychological debates, you may agree with me that the aspects to compare them can vary wildly, for example, based on their position in the political spectrum from right to left. They can be compared also by the number of years in power – four or too many – or how many wars they started, warmth in their voice, and so on.

You can also see that all these comparison categories, let us call them factors, have different units – years, meters – or no units at all! For example, there are qualities that are hard to quantify, such as the perceived warmth of the president during a speech, or projected confidence, among many others. More about leaders and followers can be found in Section 5.7.

A.6 Oscillatory baffle reactor

We can assume the "current batch" reactor to be a big vat in which a reaction takes place, but which is inefficient in many ways, such as due to poor mixing. The new solution is an oscillatory baffle reactor, "predicted OBR"; a device that restrains the flow of a fluid or gas and improves mixing, see Figure A.5. Among the claimed advantages, the OBR allows continuous processing leading to a 100-fold reduction in reactor size and greater operational control and flexibility. One of the applications where OBRs have been used is in saponification reaction systems [106].

The greatest incentive or driver for using the OBR continuous processing is safety, because the batch reactor operates with a large volume of solvent at a temperature above its boiling point at ambient pressure. Safety is improved when the operating temperature is 85 °C, closer to the ambient-pressure boiling point of the solvent. By operating in continuous mode, the inventories of solvent are lowered, improving safety too. Lower-temperature operation results in energy savings, although the reduction is primarily due to more efficient heat transfer.

There are other considerations that (chemical) engineers would love to know, such as the reduction in residence time distribution, but I refer to the article for those interested [86]. Even when Table A.2 has only four rows, you can already see that mixing all these factors to make a decision is almost impossible. Each row has different units (degrees Celsius, barg, m^3, ...) and the predicted OBR does not always seem to point in the desired direction – the pressure. Here you can practice with the OBR on how to use the IF_t method. The answer can be found right here below (Table A.3).

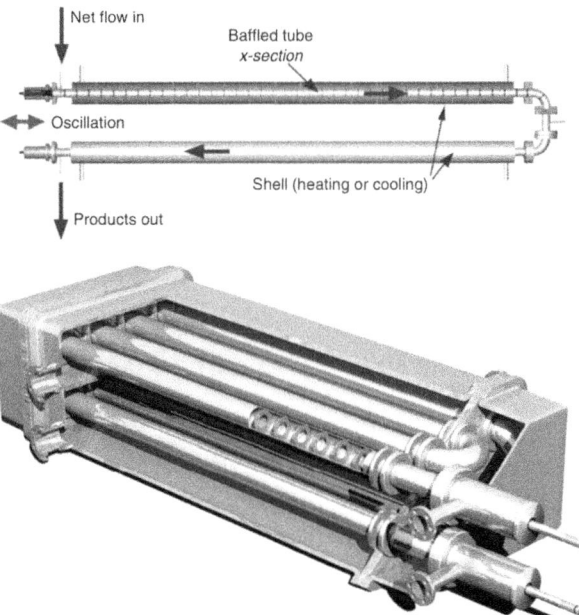

Figure A.5: The oscillatory baffle reactor is a novel form of continuous plug flow reactor. The tubes are fitted with equally spaced, low-constriction orifice plate baffles that are oscillated at low frequencies (range, 0.5 to 10 Hz). This oscillatory movement is superimposed upon the net flow of the process fluid. A conceptual design for a 500-liter industrial-scale oscillatory baffled reactor [151]. Reproduced with permission from Elsevier.

Table A.2: Test case of an oscillatory baffle reactor (OBR). Taken from [151].

Factor	Batch	OBR	d	Fraction	IF_{total}
Temperature [°C]	115	85	1	$(115/85)^1 =$??	
Pressure [barg]	1	1.7	1	$(??/??)^1 =$??	
Volume [m³]	75	0.5	1	$(??/??)^1 =$??	
Residence time [min]	120	12	1	$(??/??)^1 =$??	???

In a simplistic analysis, it is relatively easy to imagine that higher pressures can be more dangerous in the event of an accident. Therefore, the greatest incentive of improving safety is at odds. Then, even when the other rows show that the current batch reactor is not superior to the predicted OBR, how can we be sure that the OBR is the best?

This answer corresponds to the question presented just above, in Section A.6.

Table A.3: Test case of an oscillatory baffle reactor (OBR). Taken from [151].

Factor	Batch	OBR	d	Fraction	IF_{total}
Temperature [°C]	115	85	1	$(115/85)^1 = 1.35$	
Pressure [barg]	1	1.7	1	$(1/1.7)^1 = 0.59$	
Volume [m³]	75	0.5	1	$(75/0.5)^1 = 150$	
Residence time [min]	120	12	1	$(120/12)^1 = 10$	1195

The final IF is 1195 > 1, this means that the OBR reactor has an overall positive performance. We could have also added "safety" as driver, for which experts would need to assign values for each alternative, based on available experimental data or based on an arbitrary scale.

A.7 Organometallic reaction in fine chemical and pharmaceutical industry

This is the case presented for more than two alternatives in Section A.6 and presented in great measure elsewhere [86, 95]. Up to this case, this book was concerned with examples bearing only two alternatives. Here you can see three different alternatives, for which we compare 1 vs 2, 1 vs 3 and 2 vs 3.

Three different scenarios were compared for a campaign producing 5 tons of an isolated intermediate through a multistage organometallic reaction (see Figure A.6) [152].

1. The standard scenario where the reaction is performed batch-wise. Here we have six batch assets of equal size in series, each performing a specific task (protection, Li exchange, coupling, hydrolysis, extraction and distillation). The slowest step

becomes the bottleneck, which is the coupling reaction because it takes place at cryogenic temperatures to avoid side product formation.
2. A mix of continuous and batch processing, with the Li exchange and coupling reactions performed in a microreactor, making for larger costs. The reaction temperature is increased to avoid long residence times, which leads to increased overall yield (from 75 % to 80 %) and throughput increase for the coupling reaction (from 1.7 to 2.1 kg/min). The distillation is the bottleneck instead of the coupling reaction, but the workup operations (extraction, distillation, centrifugation and drying) remain the same.
3. The process synthesis design (PSD) has all reaction steps in continuous-flow operation. This gives the advantage of reducing the batch assets and the number of operators, but it requires additional investments. It is assumed that there is no further gain in yield and throughput.

The yield in all cases is preferred to be as high as possible. This is because the cost of raw material is the dominant operating cost. The next largest cost is manufacturing, and that means that the number of operators is preferred to be as small as possible. In contrast, throughput is needed as high as reachable (to decrease operating time).

A global *IF* based on those factors gives a simple indication of the reduction in operating costs, and therefore the increase in economic gain. Assuming that any additional investment, when annualized, is negligible compared to the former/conventional operating costs, when comparing the second and third alternatives against the first, it is evident that alternative 3 has a higher *IF* value (2.27). This corresponds to a better intensification of the whole change if alternative 2 would be selected (1.62).

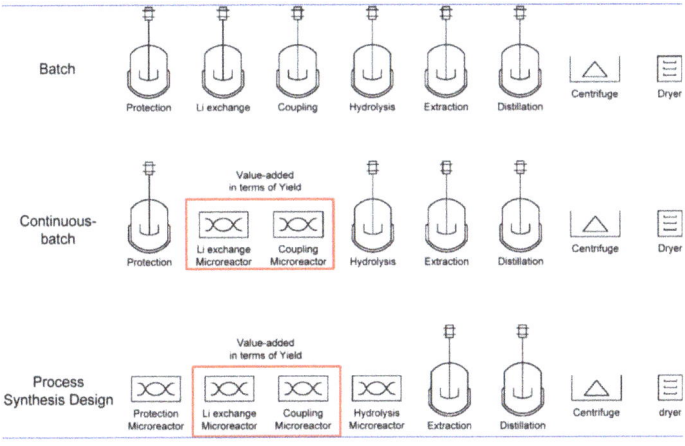

Figure A.6: Three different scenarios of commercial production using a multistage organometallic reaction: (1) batch, (2) continuous batch and (3) process synthesis design. Reprinted (adapted) with permission from [152]. Copyright 2008 American Chemical Society.

This is quantified by the 2 vs 3 calculation, where an *IF* of 1.4 is the result. Unsurprisingly, the larger *IF* value, the higher the economic gain that is finally achieved, which is in agreement with the economical gain reported by [152].

Beyond the academic exercises discussed in this book, in real-world cases the number of factors might be much higher. Consequently, the possibility of talking about single numbers (*IF*) can be much more helpful in the decision making process. Only factors with the largest impact on the chosen figure of merit should be selected and be coupled to the applicability reasoning; see Section 2.6.

Please note that the multiplicative nature of the factors F implies that after comparing case 1 vs 2 and 1 vs 3, it is not really needed to compare 2 vs 3. Thus, in practical situations such an extensive table might not be of use. For clarity purposes you can see the three situations (Table A.4).

Table A.4: Test cases for organometallic reaction [152]; assumptions and economical gain for scenarios in commercial production, assuming the current gain is 100 % for the catch case (1).

Case	Factor	B	A	d	Fraction	IF_{total}
1 vs 2	Yield gain [%]	100	105	−1	$(100/105)^{-1} = 1.05$	
	Operators [−]	3.5	2.8	1	$(3.5/2.8)^{1} = 1.25$	
	Throughput [kg/min]	1.7	2.1	−1	$(1.7/2.1)^{-1} = 1.24$	1.62
1 vs 3	Yield gain [%]	100	105	−1	$(100/105)^{-1} = 1.05$	
	Operators [−]	3.5	2	1	$(3.5/2)^{1} = 1.75$	
	Throughput [kg/min]	1.7	2.1	−1	$(1.7/2.1)^{-1} = 1.24$	2.27
2 vs 3	Yield gain [%]	105	105	−1	$(105/105)^{-1} = 1$	
	Operators [−]	2.8	2	1	$(2.8/2)^{1} = 1.4$	
	Throughput [kg/min]	2.1	2.1	−1	$(2.7/2.1)^{-1} = 1$	1.4

A.8 Answers to the rhetoric question from Section 2.3.2

Kostas Resilience, sustainability and human centricity are the main aspects of what is now dubbed industry 5.0 or the fifth industrial revolution [153].
Those can also be the robotization of workplaces and all the new challenges this may introduce when it comes to safety (physical, psychological, cognitive, etc.). You can check our publication: Redefining Safety in Light of Human-Robot Interaction: A Critical Review of Current Standards and Regulations [131].
It can also be the need for a system view on things, especially as everything now becomes part of an existing and gradually complex system or system of systems.

Akash One thing we do not currently focus on, but will be relevant in later-stage capitalism, is the social ramification. Especially in the context of aging, industrialized populations co-existing with young, unindustrialized populations. This for instance strongly affects the job redistribution.

Bibliography

[1] http://exhibits.hsl.virginia.edu/yellowfever/carlos-juan-finlay-1833-1915/, January 2022.
[2] https://david-fernandez-rivas.com/index.php/initiatives-projects/.
[3] https://www.weforum.org/agenda/2021/11/green-economy-new-technology-climate-change?utm_source=linkedin&utm_medium=social_scheduler&utm_term=Future+of+Economic+Progress&utm_content=27/11/2021+21:00.
[4] https://www.bloomberg.com/news/features/2021-10-15/rhetorical-tricks-donald-trump-and-elon-musk-use-to-control-how-you-think.
[5] https://www.weforum.org/agenda/2021/08/cognitive-bias-infographic?utm_source=linkedin&utm_medium=social_scheduler&utm_term=Education,+Gender+and+Work&utm_content=30/12/2021+23:00.
[6] https://steveblank.com/2021/12/14/i-cant-see-you-but-im-not-blind/.
[7] https://www.bakadesuyo.com/about/.
[8] https://www-technologyreview-com.cdn.ampproject.org/c/s/www.technologyreview.com/2014/10/20/169899/isaac-asimov-asks-how-do-people-get-new-ideas/amp/.
[9] https://www.weforum.org/agenda/2021/10/unilever-leena-nair-future-of-work-soft-skills-hard-skills/.
[10] https://bigthink.com/neuropsych/survival-of-the-fittest-frans-de-waal/.
[11] https://www.theatlantic.com/ideas/archive/2021/11/hot-streaks-in-your-career-dont-happen-by-accident/620514/.
[12] https://www.sciencealert.com/the-science-of-why-flow-states-feel-so-good-according-to-a-cognitive-scientist.
[13] https://davidepstein.bulletin.com/308221507559816, November 2021.
[14] https://marvelcinematicuniverse.fandom.com/wiki/Mantis, 2021.
[15] https://www.nature.com/articles/s41562-021-01217-2.
[16] https://www.macmillanihe.com/page/detail/technology-entrepreneurship-natasha-evers/?sf1=barcode&st1=9781352011173.
[17] https://www.biography.com/news/michael-phelp-perfect-body-swimming.
[18] https://www.wired.com/2001/12/aspergers/.
[19] https://www.verywellmind.com/what-are-character-strengths-4843090.
[20] https://www.viacharacter.org/topics/articles/what-are-your-signature-strengths.
[21] https://en.wikipedia.org/wiki/DIKW_pyramid/.
[22] https://www.kvk.nl/english/marketing/how-to-do-a-swot-analysis-in-5-steps/.
[23] https://www.technischwerken.nl/kennisbank/techniek-kennis/wat-betekenen-de-termen-ingenieur-engineer-engineering-en-reverse-engineering/.
[24] https://www.panoramaed.com/blog/comprehensive-guide-21st-century-skills/.
[25] https://www-technologyreview-com.cdn.ampproject.org/c/s/www.technologyreview.com/2014/10/20/169899/isaac-asimov-asks-how-do-people-get-new-ideas/amp/.
[26] https://www.innovationtraining.org/triz-training-theory-of-inventive-problem-solving/.
[27] https://theconversation.com/future-engineers-need-to-understand-their-works-human-impact-heres-how-my-classes-prepare-students-to-tackle-problems-like-climate-change-173651.
[28] https://www.nsf.gov/news/special_reports/medalofscience50/vonkarman.jsp.
[29] https://www.weforum.org/agenda/2019/09/how-to-build-an-entrepreneurial-university/.
[30] https://en.wiktionary.org/wiki/Reconstruction:Proto-Germanic/laizijan%C4%85.
[31] https://www.oxfordreference.com/view/10.1093/acref/9780191826719.001.0001/q-oro-ed4-00006236.
[32] https://ec.europa.eu/research/participants/data/ref/h2020/wp/2014_2015/annexes/h2020-wp1415-annex-g-trl_en.pdf.

https://doi.org/10.1515/9783110746822-009

[33] https://www.pdpersoneel.nl/ingenieur-technicus-of-engineer/.
[34] https://wetten.overheid.nl/BWBR0018486/2018-12-04.
[35] https://www.morganclaypool.com/doi/abs/10.2200/S00984ED1V01Y202001ETS024.
[36] https://theconversation.com/future-engineers-need-to-understand-their-works-human-impact-heres-how-my-classes-prepare-students-to-tackle-problems-like-climate-change-173651.
[37] https://www.stockholmresilience.org/research/planetary-boundaries.html.
[38] https://en.wikipedia.org/wiki/TRIZ.
[39] https://www.bbc.com/storyworks/unlocking-science/battling-bias-in-ai.
[40] https://innovationorigins.com/en/without-dialogue-every-innovation-model-is-worthless/.
[41] https://www.nature.com/articles/d41586-020-02183-x.
[42] 4TU.Centre For Engineering Education. Educating the entrepreneurial engineer, towards a 4TU special interest group.
[43] 5 exercises to help you build more empathy, March 2021. https://ideas.ted.com/5-exercises-to-help-you-build-more-empathy/.
[44] David Allen. *Getting Things Done: The Art of Stress-free Productivity*. Penguin, 2015.
[45] Ernesto Altshuler. *Guerrilla Science, Survival Strategies of a Cuban Physicist*. Springer, Cham, 2017. Number 978-3-319-51622-6.
[46] Katerina Ananiadou and Magdalean Claro. 21st century skills and competences for new millennium learners in OECD countries. OECD education working papers, no. 41. *OECD Publishing (NJ1)*, 2009.
[47] J. Andraos. *The Algebra of Organic Synthesis. Green Metrics, Design Strategy, Route Selection, and Optimization*. CRC Press-Taylor & Francis Group, Boca Raton, FL, 2012.
[48] Shane Ardo et al. Pathways to electrochemical solar-hydrogen technologies. *Energy & Environmental Science*, 11(10):2768–2783, 2018.
[49] Vishal Arghode et al. Teacher empathy and science education: A collective case study. *Eurasia Journal of Mathematics, Science and Technology Education*, 9(2):89–99, 2013.
[50] Anubhav Arora et al. Needle-free delivery of macromolecules across the skin by nanoliter-volume pulsed microjets. *Proceedings of the National Academy of Sciences*, 104(11):4255–4260, 2007.
[51] Genevieve Bell et al. Making by making strange: Defamiliarization and the design of domestic technologies. *ACM Transactions on Computer-Human Interaction (TOCHI)*, 12(2):149–173, 2005.
[52] Carla Berrospe Rodríguez et al. Toward jet injection by continuous-wave laser cavitation. *Journal of Biomedical Optics*, 22(10):105003, 2017.
[53] Rolando García Blanco. *Cien Figuras De La Ciencia En Cuba*. Nuevo Milenio, 2016.
[54] Steve Blank. *The Four Steps to the Epiphany: Successful Strategies for Products That Win*. John Wiley & Sons, 2020.
[55] Benjamin S. Bloom et al. Taxonomy of Educational Objectives. Vol. 1: Cognitive Domain. McKay, New York, 1956, 20:24.
[56] Sunday Bolade. A complementarity perspective of knowledge resources. *Journal of the Knowledge Economy*, 1–21, 2021. 10.1007/s13132-021-00743-8.
[57] Nicolas Bremond et al. Controlled multibubble surface cavitation. *Physical Review Letters*, 96(22):224501, 2006.
[58] BuBclean. BuBble bags. http://www.bubclean.nl/bubble-bags-2/.
[59] John Carreyrou. *Bad Blood: Secrets and Lies in a Silicon Valley Startup*. Alfred A. Knopf, 2018.
[60] Maureen Carroll. Stretch, dream, and do – a 21st century design thinking & STEM journey. *Journal of Research in STEM Education*, 1(1):59–70, 2015.
[61] C. Chen, M. Garedew and S. W. Sheehan. Single-step production of alcohols and paraffins from CO_2 and H_2 at metric ton scale. *ACS Energy Letters*, 7(3):988–992, 2022.

[62] Chief executives are weirder than ever. *The Economist*, November 2021.
[63] Douglas Chismar. Empathy and sympathy: The important difference. *The Journal of Value Inquiry*, 22(4):257–266, October 1988.
[64] Robert B. Cialdini. Harnessing the science of persuasion. pdf. *Harvard Business Review*, 79(72), 2001.
[65] Malissa A. Clark et al. "I feel your pain": A critical review of organizational research on empathy. *Journal of Organizational Behavior*, 40(2):166–192, 2019.
[66] James Gerald Crowther. *British Scientists of the Nineteenth Century*. Routledge, 2013.
[67] Mihaly Csikszentmihalyi. *Beyond Boredom and Anxiety*. Jossey-Bass, 2000.
[68] Katharina Cu et al. Delivery strategies for skin: Comparison of nanoliter jets, needles and topical solutions. *Annals of Biomedical Engineering*, 48:2028–2039, 2020.
[69] Tom Dalzell. *The Routledge Dictionary of Modern American Slang and Unconventional English*. Routledge, 2009. Number p. 595.
[70] Pedro De Bruyckere et al. *Urban Myths about Learning and Education*. Academic Press, 2015.
[71] Salih Emre Demirel et al. Systematic process intensification. *Current Opinion in Chemical Engineering*, 25:108–113, 2019.
[72] Erica Dhawan. *Digital Body Language: How to Build Trust and Connection, No Matter the Distance*. St. Martin's Publishing Group, 2021.
[73] A. P. Dicks and A. Hent. *Green Chemistry Metrics. A Guide to Determining and Evaluating Process Greenness*. Springer, Heidelberg, 2015.
[74] Digital Body Language for the Post-Pandemic Era. https://alum.mit.edu/slice/digital-body-language-post-pandemic-era.
[75] Nancy Duarte. *HBR Guide to Persuasive Presentations*. Harvard Business Press, 2012.
[76] M. Edwin Agwu and Henrietta N. Onwuegbuzie. Effects of international marketing environments on entrepreneurship development. *Journal of Innovation and Entrepreneurship*, 7(1):1–14, 2018.
[77] Charles E. Eesley et al. The contingent effects of top management teams on venture performance: Aligning founding team composition with innovation strategy and commercialization environment: Contingent Effects of Top Management Teams on Venture Performance. *Strategic Management Journal*, 35(12):1798–1817, December 2014.
[78] Nancy Eisenberg and Richard A. Fabes. Empathy: Conceptualization, measurement, and relation to prosocial behavior. *Motivation and Emotion*, 14(2):131–149, June 1990.
[79] Empathy: Enabling students to be entrepreneurs. https://www.thechemicalengineer.com/features/empathy-enabling-students-to-be-entrepreneurs/.
[80] Entrepreneurship for engineers, A toolbox for building a technology startup from idea to execution. https://www.edx.org/course/entrepreneurship-for-engineers.
[81] Anders Ericsson and Robert Pool. *Peak: Secrets from the New Science of Expertise*. Random House, 2016.
[82] K. Anders Ericsson, Ralf T. Krampe and Clemens Tesch-Römer. The role of deliberate practice in the acquisition of expert performance. *Psychological Review*, 100(3):363, 1993.
[83] Gerhard Ertl. Heterogeneous catalysis on the atomic scale. *The Chemical Record*, 1(1):33–45, 2001.
[84] Toni Feder. Guiding inventions from lab to market. *Physics Today*, 74(2):24–27, February 2021.
[85] Richard M. Felder. Teaching engineering at a research university. Problems and possibilities. *Educación Química*, 15(1):40, August 2018.
[86] D. Fernandez Rivas et al. Evaluation method for process intensification alternatives. *Chemical Engineering and Processing – Process Intensification*, 123:221–232, January 2018.
[87] D. Fernandez Rivas and J. G. E. Gardeniers. On the resilience of PDMS microchannels after violent optical breakdown microbubble cavitation. *ASME Conference Proceedings*, (48345):1939–1942, 2008.

[88] David Fernandez Rivas. *Taming Acoustic Cavitation*. PhD thesis, University of Twente, 9789036534192, October 2012.
[89] David Fernandez Rivas. Small bubbles and bubble bags: A scientific knowledge valorisation. *The Cuban Scientist*, 1(1):23–24, 2020.
[90] David Fernandez Rivas et al. Efficient sonochemistry through microbubbles generated with micromachined surfaces. *Angewandte Chemie International Edition*, 49(50):9699–9701, 2010.
[91] David Fernandez Rivas et al. Merging microfluidics and sonochemistry: Towards greener and more efficient micro-sono-reactors. *Chemical Communications*, 48(89):10935–10947, 2012.
[92] David Fernandez Rivas et al. Ultrasound artificially nucleated bubbles and their sonochemical radical production. *Ultrasonics Sonochemistry*, 20(1):510–524, 2013.
[93] David Fernandez Rivas et al. Process intensification education contributes to sustainable development goals. Part 1. *Education for Chemical Engineers*, 32:1–14, July 2020.
[94] David Fernandez Rivas et al. Process intensification education contributes to sustainable development goals. Part 2. *Education for Chemical Engineers*, 32:15–24, 2020.
[95] David Fernandez Rivas and Pedro Cintas. On an intensification factor for green chemistry and engineering: The value of an operationally simple decision-making tool in process assessment. *Sustainable Chemistry and Pharmacy*, 27:100651, 2022. 10.1016/j.scp.2022.100651.
[96] David Fernandez Rivas and Sebastian Husein. Empathy, Persuasiveness and Knowledge promote innovative engineering and entrepreneurial skills. *Education for Chemical Engineers*, 2022. 10.1016/j.ece.2022.05.002. https://www.sciencedirect.com/science/article/pii/S1749772822000161.
[97] David Fernandez Rivas and Simon Kuhn. Synergy of microfluidics and ultrasound: Process intensification challenges and opportunities. *Topics in Current Chemistry (Cham)*, 374(5):70, 2016 Oct. doi:10.1007/s41061-016-0070-y. Epub 2016 Sep 21. PMID: 27654863; PMCID: PMC5480412.
[98] David Fernandez Rivas and B. Verhaagen. Preface to the special issue: Cleaning with bubbles. *Ultrasonics Sonochemistry*, 29:517–518, 2016.
[99] Richard P. Feynman et al. The Feynman lectures on physics; vol. I. *American Journal of Physics*, 33(9):750–752, 1965.
[100] Samuel Fleischacker. *Being Me Being You: Adam Smith and Empathy*. University of Chicago Press, 2019.
[101] Thomas L. Friedman. *Thank You for Being Late: An Optimist's Guide to Thriving in the Age of Accelerations (version 2.0, with a New Afterword)*. Picador USA, 2017.
[102] Javier Garcia-Martinez. Chemistry 2030: A roadmap for a new decade. *Angewandte Chemie International Edition*, 60(10):4956–4960, 2021.
[103] Javier García-Martínez and Kunhao Li. *Chemistry Entrepreneurship*. Wiley-VCH, 2021.
[104] John Gribbin. *The Scientists: A History of Science Told through the Lives of Its Greatest Inventors*. Random House Trade Paperbacks, 2004. ISBN 9780812967883.
[105] James Gribble and Graham Oliver. Empathy and education. *Studies in Philosophy and Education*, 8(1):3–29, March 1973.
[106] A. P. Harvey et al. Operation and optimization of an oscillatory flow continuous reactor. *Industrial & Engineering Chemistry Research*, 40(23):5371–5377, 2001.
[107] Byron Hempel et al. Scalable and practical teaching practices faculty can deploy to increase retention: A faculty cookbook for increasing student success. *Education for Chemical Engineers*, 2020.
[108] S. G. Hernandez and S. W. Sheehan. Comparison of carbon sequestration efficacy between artificial photosynthetic carbon dioxide conversion and timberland reforestation. *MRS Energy & Sustainability*, 7:E32, 2020.

[109] S. Hodges et al. Empathy: Encyclopedia of Social Psychology, 2007.
[110] Davide Iannuzzi. *Entrepreneurship for Physicists: A Practical Guide to Move Inventions from University to Market*. Morgan & Claypool Publishers, 2017.
[111] Sheila Jasanoff. Technologies of humility. *Nature*, 450(7166):33, 2007.
[112] D. Jeffrey. Communicating with a human voice: Developing a relational model of empathy. *Journal of the Royal College of Physicians of Edinburgh*, 47(3):266–270, 2018.
[113] D. Jeffrey and R. Downie. Empathy – can it be taught? *Journal of the Royal College of Physicians of Edinburgh*, 46(2):107–112, 2016.
[114] Aldert Kamp. Engineering education in the rapidly changing world: Rethinking the vision for higher engineering education, 2016.
[115] Aldert Kamp. Science & technology education for 21st century Europe. Technical report, 2019.
[116] Aldert Kamp. Navigating the landscape of higher engineering education, 2020.
[117] Steven N. Kaplan and Morten Sorensen. Are CEOs Different? SSRN Scholarly Paper ID 2747691, Rochester, NY, July 2020.
[118] Paul A. Kirschner and Carl Hendrick. *How Learning Happens: Seminal Works in Educational Psychology and What They Mean in Practice*. Routledge, 2020.
[119] Lauren Kopajtic. Review of being me being you: Adam smith and empathy. Online https://ndpr.nd.edu/reviews/being-me-being-you-adam-smith-and-empathy/, 2020.
[120] Fred Aj Korthagen and Angelo Vasalos. Going to the core: Deepening reflection by connecting the person to the profession. In *Handbook of Reflection and Reflective Inquiry*, pages 529–552. Springer, 2010.
[121] Denise Krch. *Cognitive Processing*. Springer New York, New York, NY, 2011.
[122] A. Lapkin and D. J. C. Constable. *Green Chemistry Metrics. Measuring and Monitoring Sustainable Processes*. John Wiley & Sons, 2009.
[123] Letter to You, November 2021. Page Version ID: 1056141741. https://en.wikipedia.org/w/index.php?title=Letter_to_You&oldid=1056141741.
[124] Bingjie Liu and S. Shyam Sundar. Should machines express sympathy and empathy? Experiments with a health advice chatbot. *Cyberpsychology, Behavior, and Social Networking*, 21(10):625–636, October 2018.
[125] Paul Lockhart. *Measurement*. Harvard University Press, Cambridge, 2014, p. 2.
[126] Detlef Lohse et al. Sonoluminescing air bubbles rectify argon. *Physical Review Letters*, 78(7):1359, 1997.
[127] Detlef Lohse. Bubble puzzles: From fundamentals to applications. *Physical Review Fluids*, 3(11):110504, 2018.
[128] G. Maarten Bonnema et al. *Systems Design and Engineering: Facilitating Multidisciplinary Development Projects*. CRC Press, 2016.
[129] Magalí Mercuri and David Fernandez Rivas. Challenges and opportunities for small volumes delivery into the skin. *Biomicrofluidics*, 15(1):011301, 2021.
[130] Franco Malerba and Maureen Mckelvey. Knowledge-intensive innovative entrepreneurship integrating Schumpeter, evolutionary economics, and innovation systems. *Small Business Economics*, 54(2):503–522, 2020.
[131] Alberto Martinetti, Peter K. Chemweno, Kostas Nizamis and Eduard Fosch-Villaronga. Redefining safety in light of human-robot interaction: A critical review of current standards and regulations. *Frontiers in Chemical Engineering*, 3:666237, 2021.
[132] Javier Garcia Martinez. Javier García-Martínez on the commercialization of research.
[133] James Clerk Maxwell. *The Scientific Papers of James Clerk Maxwell..., vol. 2*. University Press, 1890.
[134] Albert Mehrabian and Norman Epstein. A measure of emotional empathy. *Journal of Personality*, 40(4):525–543, 1972. eprint: https://onlinelibrary.wiley.com/doi/pdf/10.1111/j.1467-6494.1972.tb00078.x.

[135] Theo Melrose. Improve at anything with deliberate practice; how deliberate practice leads to success, 2022.
[136] Miquel A. Modestino et al. The potential for microfluidics in electrochemical energy systems. *Energy & Environmental Science*, 9(11):3381–3391, 2016.
[137] N. a. Scanning the environment: PESTEL analysis. https://www.business-to-you.com/scanning-the-environment-pestel-analysis/.
[138] J. Ohm et al. Covid-19: Women, equity, and inclusion in the future of work (report). https://www.catalyst.org/research/covid-report-workplace-women/.
[139] Seun Azeez Olugbola. Exploring entrepreneurial readiness of youth and startup success components: Entrepreneurship training as a moderator. *Journal of Innovation & Knowledge*, 2(3):155–171, September 2017.
[140] Charlotte Oude Alink and Hans Van Den Berg. Project-led education. https://www.utwente.nl/en/ces/celt/publications/20130820-ple-final.pdf.
[141] Own the room. https://disneyplusoriginals.disney.com/movie/own-the-room.
[142] Loreto Oyarte Gálvez et al. Microfluidics control the ballistic energy of thermocavitation liquid jets for needle-free injections. *Journal of Applied Physics*, 127(10):104901, 2020.
[143] G. Pace and S. W. Sheehan. Scaling CO_2 capture with downstream flow CO_2 conversion to ethanol. *Frontiers in Climate*, 3:35, 2021.
[144] J. P. Padilla-Martinez, C. Berrospe-Rodriguez, Guillermo Aguilar, J. C. Ramirez-San-Juan, and Ruben Ramos-Garcia. Optic cavitation with CW lasers: A review. *Physics of Fluids*, 26(12):122007, 2014.
[145] Maria Pappaterra et al. Cavitation intensifying bags improve ultrasonic advanced oxidation with Pd/Al_2O_3 catalyst. *Ultrasonics Sonochemistry*, 70:105324, 2021.
[146] Iulian Patraşcu et al. Eco-efficient downstream processing of biobutanol by enhanced process intensification and integration. *ACS Sustainable Chemistry & Engineering*, 6(4):5452–5461, 2018.
[147] Reidar Pedersen. Empathy development in medical education – A critical review. *Medical Teacher*, 32(7):593–600, July 2010. 10.3109/01421590903544702.
[148] Luis Perez-Breva. *Innovating: A Doer's Manifesto for Starting from a Hunch, Prototyping Problems, Scaling Up, and Learning to Be Productively Wrong*. MIT Press, Cambridge, MA, 2016.
[149] Miguel A. Quetzeri-Santiago et al. Impact of a microfluidic jet on a pendant droplet. *Soft Matter*, 17(32):7466–7475, 2021.
[150] Lubna Rashid. Entrepreneurship education and sustainable development goals: A literature review and a closer look at fragile states and technology-enabled approaches. *Sustainability*, 11(19), 2019.
[151] David Reay et al. *Process Intensification: Engineering for Efficiency, Sustainability and Flexibility*. Butterworth-Heinemann, 2013.
[152] Dominique M. Roberge et al. Microreactor technology and continuous processes in the fine chemical and pharmaceutical industry: Is the revolution underway? *Organic Process Research & Development*, 12(5):905–910, 2008.
[153] David Romero and Johan Stahre. Towards the Resilient Operator 5.0: The future of work in smart resilient manufacturing systems. *Procedia CIRP*, 104:1089–1094, 2021.
[154] Philip T. Roundy and Thomas S. Lyons. Humility in social entrepreneurs and its implications for social impact entrepreneurial ecosystems. *Journal of Business Venturing Insights*, 17:e00296, 2022.
[155] Harri Ruoslahti. Complexity in project co-creation of knowledge for innovation. *Journal of Innovation & Knowledge*, 5(4):228–235, October 2020.
[156] Aidin Salamzadeh and Hiroko Kawamorita Kesim. Startup companies: Life cycle and challenges. In *4th International Conference on Employment, Education and Entrepreneurship (EEE)*, Belgrade, Serbia, 2015.

[157] Jelle Schoppink and David Fernández Rivas. Jet injectors: Perspectives for small volume delivery with lasers. *Advanced Drug Delivery Reviews*, 182:114109, 2022.
[158] Joseph A. Schumpeter. The Theory of Economic Development: An Inquiry into Profits, Capital, Credit, Interest, and the Business Cycle (1912/1934). Transaction Publishers, January 1982, 1:244.
[159] Joseph A. Schumpeter. *Economic Theory and Entrepreneurial History*. Harvard University Press, 2013.
[160] Sophie Sitter et al. An overview of process intensification methods. *Current Opinion in Chemical Engineering*, 25:87–94, 2019.
[161] Adam Smith. *The Theory of Moral Sentiments, vol. 1*. I, Richardson, 1022.
[162] Adam Smith. *The Wealth of Nations [1776]*, volume 11937. na, 1937.
[163] Medadi E. Ssentanda. The challenges of teaching reading in Uganda: Curriculum guidelines and language policy viewed from the classroom. *Apples: Journal of Applied Language Studies*, 8(2):1–22, 2014.
[164] Nassim Nicholas Taleb. *Skin in the Game: Hidden Asymmetries in Daily Life*. Penguin Books Ltd, 2019.
[165] Melanie Tervalon and Jann Murray-Garcia. Cultural humility versus cultural competence: A critical distinction in defining physician training outcomes in multicultural education. *Journal of Health Care for the Poor and Underserved*, 9(2):117–125, 1998.
[166] Sharon Tettegah and Carolyn J. Anderson. Pre-service teachers' empathy and cognitions: Statistical analysis of text data by graphical models. *Contemporary Educational Psychology*, 32(1):48–82, 2007.
[167] Owen Thomas. Fake it until you make it: A silicon valley strategy that seems unstoppable.
[168] H. Holden Thorp. It's not as easy as it looks. *Science*, 374(6575):1537, 2021.
[169] Akira Utsumi. Toward a cognitive model of poetic effects in figurative language. In *IEEE International Conference on Systems, Man and Cybernetics*, vol. 7, 6 pp. IEEE, 2002.
[170] Ana M. Valenzuela-Toro and Mariana Viglino. How Latin American researchers suffer in science. *Nature*, 598(7880):374–375, September 2021.
[171] Tara Van Bommel. The power of empathy in times of crisis and beyond. https://www.catalyst.org/reports/empathy-work-strategy-crisis/.
[172] Tom Van Gerven and Andrzej Stankiewicz. Structure, energy, synergy, time: The fundamentals of process intensification. *Industrial & Engineering Chemistry Research*, 48(5):2465–2474, March 2009.
[173] B. Verhaagen et al. Micropits for ultrasonic treatment. Technical report, 2015.
[174] Bram Verhaagen and David Fernández Rivas. Measuring cavitation and its cleaning effect. *Ultrasonics Sonochemistry*, 29:619–628, 2016.
[175] Claas Willem Visser et al. In-air microfluidics enables rapid fabrication of emulsions, suspensions, and 3D modular (bio)materials. *Science Advances*, 4(1):eaao1175, 2018.
[176] Jyrki Wallenius et al. Multiple criteria decision making, multiattribute utility theory: Recent accomplishments and what lies ahead. *Management Science*, 54(7):1336–1349, 2008.
[177] Timothy G. Walmsley et al. Frontiers in process development, integration and intensification for circular life cycles and reduced emissions. *Journal of Cleaner Production*, 201:178–191, 2018.
[178] Robert S. Weber et al. The scaling economics of small unit operations. *Journal of Advanced Manufacturing and Processing*, 3(1):e10074, 2021.
[179] Dianne H. B. Welsh et al. Entrepreneurship education: Process, method, or both? *Journal of Innovation & Knowledge*, 1(3):125–132, September 2016.
[180] Geoffrey West. *Scale: The Universal Laws of Life, Growth, and Death in Organisms, Cities, and Companies*. Penguin, 2018.

[181] Chalen Westaby and Emma Jones. Empathy: An essential element of legal practice or 'never the twain shall meet'? *International Journal of the Legal Profession*, 25(1):107–124, January 2018. 10.1080/09695958.2017.1359615.
[182] Carl Wieman. Applying new research to improve science education. *Issues in Science and Technology*, 29(1):25–32, 2012.
[183] Jamil Zaki and Roach Cydney. *Leading with Empathy in Turbulent Times: A Practical Guide*. Edelman, 2021.

Index

21st Century Skills 71

ALACT 71
Albert Einstein 91
Alfo J. Batista Leyva 25
applicability 25, 32

Bas Koelewijn 15
bias 94
BuBble Gun 58
bubbles 10
DuDLlean 52

Carlos Juan Finlay de Barres 20
CBL 69
company 52, 61, 77
Connie Nshemereirwe 125

Daniela Blanco 121
Detlef Lohse 50
durable skills 11, 70, 119

E_3 6
Elizabeth Holmes 107
empathic entrepreneurial engineers 47
empathy 13, 39
energy 43
engineer 6, 64
engineering 88
entrepreneur 8

failure 113
FlowBeams 61
follower 104

Getting things done (GTD) 102

Han Gardeniers 6
Harold "Doc" Edgerton 17
humility 68
hunch 82, 85

IF 41, 54
ingenuity 86
innovation 8, 37, 68, 86, 96, 112
intensification factor 27, 32
Isaac Asimov 104

J. W. Gibbs 7
Jim Heirbaut 14
John Dalton 36

knowledge 9, 39
KPE XVIII, 1, 3, 70
KPE ratios 40

leader 104
learning objectives 62
Lord Kelvin 25

Marike ter Maat 73
Miguel A. Modestino 123

onion 103

persuasiveness 11, 39
PESTEL 38
propaganda 90

Richard Novak 127
Robert Lepenies 14

science 7
science, technology and society 7
SEST 44
society 7
soft skills 11
Stafford W. Sheehan 129
structure 43
success 113
Sustainability, Innovation, Diversity and
    Education (SIDE) 65
SWOT 38
sympathy 15
synergy 43

team 111
technologies of humility 69
Theodore von Kármán 7
time 43, 114
tinkering 82, 93
Tom Kamperman 133
top dog 110
TRIZ 85
tunnel vision 33

uncertainty 69
United Nations Sustainable Development Goals
    (UNSDG) 42

WINE 101
WOOP 105

https://doi.org/10.1515/9783110746822-010

www.ingramcontent.com/pod-product-compliance
Lightning Source LLC
Chambersburg PA
CBHW080937300426
44115CB00017B/2859